The Love Sense

Love Isn't What You Think:
The Case for Love as Our Sixth Sense

L. Milena Powell

Publishing Notes:

ISBN: 978-0-9977496-9-4
All text written by L. Milena Powell
RUMI translation public domain
Published by: StarRockGirl to New Earth Productions, USA
1st Edition February 2026
Printed by: KDP

TABLE OF CONTENTS

Love is the astrolabe of God's mysteries

— *Rumi,* Masnavi, Book I

INTRODUCTION
The Search for What We Cannot Name

I didn't set out to question the nature of love, but love found me anyway, like a thread I couldn't stop pulling. For as long as I can remember, I've been in search of something I *had* to understand. It wasn't a partner or success. It wasn't even life's meaning, though meaning threaded itself through the journey. I was seeking the thread that ran beneath every spiritual inquiry, every mystical experience, every miracle I ever had. I was searching for consciousness itself — not the kind we talk about in books or conferences, but the kind that lives behind the veil of form and breathes us into being, that vibrates at the edges of what we call reality. It remembers. It *knows*.

Like many seekers, I was guided by questions I didn't have answers for. How does consciousness work? Where does it reside and what is it made of? How does it preexist form entirely? And more pressingly, what am I, really?

I started my seeking through science — a degree in biochemistry to figure out what I am and how I exist. What I learned was mostly limited to our physical bodies, and ultimately, that every single thing that keeps our body functioning is a chemical reaction or a physics phenomenon, and the chemical reactions are just physics too, causing molecular shape and particle charge or waveform interactions. The benefit was that it introduced me to physics, where I hoped to find more—and did, a bit. But I decided to leave physics to the experts as that dead-ended into mathematics and formulas. Instead I turned inward, seeking something more experiential, in meditative and post/deep meditative states.

As I kept pulling at the thread, I hit an invisible wall, conceivably due to the limitation built into our biology. We are, after all, beings of five senses, trained to perceive the visible spectrum, the audible range, the measurable pattern. (You can go a little deeper in meditation and remote viewing, where quantum science starts to appear.) We build our truths from what we can touch, test, and repeat. I concluded early on that the structure of consciousness — our own and the universe's — must simply be beyond the reach of these senses.

What if, like a fish in water, we are swimming in dimensions we cannot perceive, trying to define something we're immersed in but cannot see? Beyond my meditations, I also began to contemplate not only what consciousness is, but how we might detect it if we can't directly perceive it. If a multidimensional self were intersecting this world, what trace would it leave behind? What shape? What signal? What sensation?

That's when I found love — not romantic love or an idealized or sentimental love, but a very specific sensation — a vibrational state that kept showing up as soon as I touched something real. It felt like home. I realized I wasn't just searching for consciousness. I was also searching for love, and the two were not separate. They were reflections of the same frequency, a dimensional echo of the Source I come from, and the one we all belong to. This book is the map of that realization.

What If Love Is a Sense?

This book was first seeded during my 2017 Poster Presentation at the Science and Nonduality (SAND) Conference, where I introduced the hypothesis that love is our sixth sense. That presentation, a visual and conceptual exploration into the nature of love as a perceptual faculty rather than an emotion, planted the foundation for what would become this larger inquiry.

At the time, I proposed that our persistent search for love is not merely psychological or cultural, but rather a deeply encoded instinct pointing toward our multidimensional nature. This book is the evolution of that idea. It takes the core hypothesis from the poster board into the living, breathing, spiraling terrain of direct inquiry, expanded metaphysics, and experiential consciousness, unfolding the question I began asking years ago:

What if love is the sensory trace of a higher-dimensional self?

For centuries, science has approached love as either a neurochemical reaction including dopamine, oxytocin, vasopressin, or as an evolutionary strategy for bonding and reproduction.

Psychology frames it as attachment. Sociology calls it a cultural construct. While these models offer insight, they fail to account for the most profound experiences of love, the kind that arises in deep meditation, in the presence of an enlightened being, during Reiki or energy work, or in the simple, unexplainable moment when the heart opens to everything.

How do we explain unconditional love — a state that transcends the body, the mind, and the need for an object? How do we explain the feeling of being held by something vast and benevolent, even in solitude? How do we explain the resonance between two beings who have never met, or the palpable field that arises when a group meditates together? Science has no answer for this. Not yet.

What if the answer has been hiding in plain sight? What if love is not something we produce, but something we perceive? What if, just like sight and sound, love is a sensory modality — a faculty by which the body-mind detects and responds to an energetic stimulus in the environment? What if love is a sense?

The Thesis

This book proposes a radical rethinking of love, not as an emotion, not as a biochemical response, and not as a psychological construct, but as a sense — a perceptual system that allows us to detect and attune to the vibrational frequency of Source, of unity, of the quantum field itself. This assumes consciousness is fundamental, something I have always believed through scientific exploration, personal experience, and also perhaps, a greater remembering.

Consider the definition of a sense: A faculty by which the body perceives an external stimulus. Our five senses work by translating energy into electrical signals the brain can interpret. Light waves become sight. Sound waves become hearing. Chemical molecules become taste and smell. Pressure becomes touch.

What about love? We feel it. We know it. It changes our physiology including our heart rate, our breathing, our neural activity. It has a presence. And yet, we don't see it, hear it, taste it, smell it, or touch it in the way we do other stimuli. So how do we perceive it?

This book will argue that love is sensed vibrationally — through the body's bioelectric and biomagnetic field, through quantum coherence in our cells, through the heart's electromagnetic intelligence, and through a subtle, often overlooked faculty that connects us to higher-dimensional reality.

We will explore:

- How the body senses love — through dopamine pathways, the limbic system, and the D3 receptor, and also through bioelectric fields, interoception, and quantum processes science is only beginning to map.

- What defines a sense — and why love meets every criterion.

- The vibrational nature of reality — how quantum physics, dimensional theory, and consciousness science converge to suggest that love is not in us, but that we are in love — immersed in a field we can tune into.

- Why we search for love relentlessly — because it is the compass guiding us back to Source, to unity, to the higher-dimensional self we truly are.

- What happens when we treat love as a sense and how it changes relationships, healing, meditation, and the trajectory of human evolution.

How to Read This Book

This is not a deeply scientific text, though science threads through every chapter. It is not a spiritual manual, though mysticism and direct experience inform the inquiry. It is a bridge between the measurable and the ineffable, between the brain and the heart, between the particle and the wave.

If you are a scientist, you will find research here — neurobiology, quantum coherence, plant bioacoustics, biofield studies — that invites you to reconsider the limits of materialism.

If you are a mystic, you will find validation here — that what you have always known is beginning to show up in the data.

If you are simply someone who has loved deeply, meditated quietly, or felt the presence of something greater than yourself, you will find a framework here that makes sense of what you have experienced.

This book is organized into four parts:

Part I: The Search — Why everyday humans are on a relentless quest for love, and why all spiritual paths seem to lead there.

Part II: The Case for Love as a Sense — What defines a sense, how love fits the criteria, and the biology, physics, and consciousness science that support this hypothesis.

Part III: Consciousness and the Field — How love functions as a dimensional bridge, a quantum resonance, and a vibrational attunement to Source.

Part IV: Evolution and Awakening — What changes when we recognize love as a sense, and how it might guide humanity's next evolutionary leap.

Each chapter builds on the last, weaving together neuroscience, quantum physics, plant intelligence, mystical experience, and the direct, lived knowing of those who have touched the field of love and returned changed.

An Invitation To Consider Something New

If you are a scientist, this book asks only that you suspend any initial disbelief and never abandon science, but potentially expand it. This theory reorganizes what we already know. Also to consider that the senses we know, may not be all we have and that the love we seek may not be something we must earn, find, or manufacture, but something we are already immersed in — a field we can learn to perceive more clearly.

If my hypothesis is true, if love is indeed a sense, then the implications are vast. It means we are not isolated beings in a cold, mechanical universe. It means we are perceptual instruments tuned to a frequency most of us have forgotten how to hear. It means the search for love is not a neurosis or a weakness, but a navigational instinct, and the body-mind's way of orienting itself toward home.

It means that the more we learn to sense love in ourselves, in others, in the world, in the silence of meditation, in the hum of the quantum field, the more we remember what we truly are.

Welcome to The Love Sense.

CHAPTER ONE
The U-Shaped Trace - What Higher Dimensions Leave Behind

My inquiry started, as many strange things do, with a gif. I was mesmerized, and would not stop staring. A friend had posted it on social media — a simple animation of a metal rod, held at an angle, rotating around a central axis. It's called a hyperbolic slot. The rod itself was straight, a solid, three-dimensional object, but as it passed through a flat, two-dimensional plane of plexiglass, something unexpected happened. The rod didn't appear as a rod. It didn't even appear as a straight line. Instead, it carved out a wide U-shape, a curved, bending trace that made no sense in the logic of the flat plexiglass world it was passing through.

I watched it loop. Again. And again. I couldn't look away. I came back to it for days. The two-dimensional plane could only perceive a slice of the rod — the part of it that intersected its reality at any given moment. To the beings living in that flat world (if there had been any), the rod would seem mysterious, fragmented, impossible to fully grasp. They could see the cut out wide U-shape forming, and an oval entering and disappearing, but they could never see the whole — the simple, straight rod rotating above them, beyond the limits of their perception.

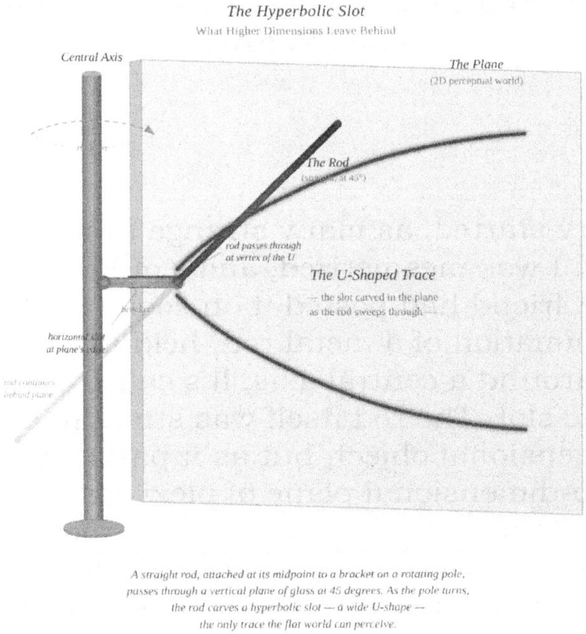

Figure 1: The Hyperbolic Slot — a straight rod passing through a flat plane leaves only a curved trace behind.

To two dimensional viewers, the cut out U-shape would be the only reality they knew. They might study it. Measure it. Build theories around it. Yet no matter what they did, they would always be looking at a trace — a shadow of something larger, something their perceptions, their senses were not designed to detect.

And that's when it hit me about my missing information from my life long consciousness inquiry. We are those beings. We too are living in a limited perceptual plane, perceiving only traces of higher-dimensional realities passing through us, intersecting our world in ways we can sense but not fully see. I was lost in thought for days or weeks or even months, pondering the potentiality of the many (or few) things we may be able to see from our limited human viewpoint of our multidimensional nature, before it hit me — what if love, something I experienced fully in my deep meditative states, is one of those traces?

The Flatland Problem

The idea isn't new. In 1884, Edwin Abbott wrote Flatland: A Romance of Many Dimensions, a satirical novella that imagined a two-dimensional world where beings could only perceive length and width — no height, no depth. When a three-dimensional sphere passed through their world, they saw it as a circle that appeared, grew, shrank, and disappeared, never understanding that what they were seeing was a slice of something greater.

The Flatlanders could measure the circle. They could track its movement. They could even predict its behavior. But they could never see the sphere — not because it wasn't there, but because their senses were limited to two dimensions. We chuckle at the Flatlanders. We think, Of course they can't see the sphere — they're stuck in 2D! But are we so different? Most physicists and theoreticians believe that all dimensions likely co-exist in the same space-time construct. We simply can't perceive them with our third-dimensional senses.

Our eyes see only 0.0035% of the electromagnetic spectrum. Our ears hear only a narrow band of sound frequencies. Our touch, taste, and smell are optimized for survival in this density, but they miss so much. So the question becomes, "What are we missing?"

If higher-dimensional objects, energies, or consciousnesses intersect our reality, what would we perceive? Not the whole — we can't. But we could see the trace, the slice, the U-shaped cutout left behind. I made a list of possibilities and negated most of it. Yet my daily meditations and subsequently society's obsession with love, kept me coming back.

What if, among all the things we cannot fully see, among all the subtle energies, quantum coherences, and vibrational fields modern science is only beginning to map, love ends up being one of the higher dimensional traces? What if love is not something we produce, but something we detect — a higher-dimensional frequency intersecting our biology, leaving a trace we can feel and the body can process, but one we still do not fully comprehend?

The Invisible Shape of Reality

This wide U-shaped imprint becomes our key metaphor. It is not the rod itself, nor the totality of its movement that we need to address, only what can be seen from within a limited frame of perception. In our metaphor, the rod does not announce itself; it offers no evidence of its rotation, its length, or even its true dimensionality unless one is able to step outside the plane. But for those of us inside the plane, locked into a sensory system evolved for survival in three-dimensional space, such a leap is not straightforward. We are creatures within the plexiglass, trying to map the motion of the rod by watching only the brief intersection of its diameter, and the bending cuts it our world that it leaves behind, the sole action of its presence.

This is how I came to see love — not as a feeling. Not even, primarily, as a connection to another. But as a U-shaped trace, a bending in the plane of human life left behind by something higher-dimensional as it moves through us.

The longing we feel for love is not, at its root, a longing for a person, or even for a return to innocence. It is the longing to perceive what is always passing through us but remains partially invisible. And of all the things we chase such as joy, power, safety, it is love that most consistently reveals itself as the unshakable pursuit. Why? Because love is the signature the rod leaves in the glass. That signature is not accidental. It is precise, patterned, and repeatable. This means it can be studied, mapped, and perhaps even understood as the sensory language of the multidimensional.

This is the critical shift. If we stop thinking of love as a human emotion and begin to understand it as a dimensional artifact — the shadow-trace of a greater geometry of consciousness — we begin to ask new questions.

We no longer think "Why do I always fall for unavailable partners?" but rather, "What is my being attempting to remember through this gravitational pull toward the unseen?" The grief of unrequited love, the surge of recognition in a stranger's eyes, the peace of unconditional presence are not fluctuations in mood. They are intersections — points at which a larger field passes through our perceptual plane.

And just like the U-shape on the plexiglass, love appears distorted because it's angled, and that is because it's not from here. It's distortion is a clue, not a failure. Its curve is a whisper of a whole we cannot yet grasp.

What We're Looking For

Let us pause and ask, what would make a species evolve with such a deep-seated hunger for love that it becomes the driving force behind most decisions, sacrifices, creations, and collapses? Is it really just hormonal? Why would evolution, if purely mechanical and survival-based, risk the inefficiency of such vulnerability?

The answer may be that love is not inefficient. It is instructional. It teaches not by doctrine but by direct resonance. It tunes the human instrument toward something greater, even when we don't understand the source of the melody. In this way, the longing for love, especially the ache of its absence, is not a weakness of the psyche, but an echo of memory. Perhaps it's a faint trace of a memory of something we once touched, or were once made entirely of — and in truth, still are.

Just as the rotating rod never leaves the space above the plexiglass plane, yet constantly intersects it, love may be a constant, passing through us in fragments that appear episodic, ecstatic, or elusive. The illusion is that love comes and goes. The deeper truth is that we move in and out of alignment with it, like a plane catching or missing the angled rod's trace. This begins to offer a functional definition:

Love is the perceptible trace left by the intersection of higher-dimensional coherence with human awareness.

That trace is not always soft. Sometimes it erupts as heartbreak, spiritual crisis, or a breaking open of identity. These, too, are resonant effects of a higher structure encountering our limited capacity. They do not mean something has gone wrong. They mean something larger has made contact.

And yet, human culture often treats love as anomaly, an unpredictable gift or dangerous indulgence. We write poetry about it. We fear its loss. We seek its permanence in impermanent places. But what if all of these are attempts to describe, stabilize, or prolong our contact with a multidimensional truth that cannot fully land in the current shape of our nervous systems?

From this lens, our task becomes not to possess love, but to attune to its geometry and learn to recognize its trace when it arrives, to stop mistaking the person, event, or moment that carried it for the source itself. *To perceive love as a field signature, not a possession.* And to become increasingly available to that signature in ourselves, in others, and in the greater architecture of being.

This reframes the question. Not "How do I find love?" but "*What part of me is able to recognize love's multidimensional imprint when it arises?*"

The Shape Made Visible

The wide U-shaped trace is not the thing itself. It is a shadow of an angle, a whisper of a higher movement translated downward. So too with love. We may encounter it in a glance, in the trembling intimacy of connection, or in the ache of its apparent withdrawal, but these are only expressions, curved shadows , of something larger brushing against our perception.

Love, in this sense, is not a feeling but a field. A trans-dimensional curve passing through us. A resonance that cannot be fully held in a linear mind. When we touch it, we remember what we are. When it passes, we often forget.

This is why the longing persists. Not as neurosis, but as recognition. Not as deficiency, but as directional gravity.

If we take this seriously, then the implications are profound. It suggests that human consciousness is not bounded by the five senses but tuned by what is beyond them. That the ache for love is a navigational tool, a built-in mechanism for aligning with multidimensional memory. That this memory contains not just personal desire, but the architecture of Being itself. Thus, love is both the call and the compass. The shape and the signal.

And if love is the shape made visible as higher-dimensional coherence intersects our awareness, then learning to perceive it more clearly is not a sentimental pursuit. It is an act of dimensional alignment. It is an initiation into consciousness.

In the chapters to come, we will explore this alignment through the metaphor of geometry, through scientific inquiry, through ancient wisdom, and through the subtleties of human perception. We will follow the trace of the rod. We will look more closely at the plexiglass. We will ask, what else leaves a pattern on this plane of our own perceptions?

But we begin here, with the premise that love is not a mere emotion. It is evidence of something larger we are part of, something coherent enough to pull us home, to our conscious evolution.

CHAPTER TWO
The Search for What Bleeds Through

So how did I arrive at the feet of love? For months, I couldn't shake the gif, and found myself returning to a singular, haunting question, "What part of our vast, multidimensional nature manages to penetrate into our ordinary human awareness that we can't quite see clearly"?

Our multidimensional consciousness — the fullness of what we are as expressions of Source — exists in dimensions our human perception simply cannot encompass. We are three-dimensional beings trying to comprehend a reality that extends far beyond the constraints of height, width, and depth. Most of what we are remains invisible to us, like the bulk of that sphere existing beyond Flatland's plane. *Many* things should bleed through. Something from that vast, incomprehensible wholeness manages to seep into our limited awareness, leaving traces we can detect, measure, and, if we're paying attention, can recognize. I became obsessed with identifying what that something was.

The Criteria

I knew whatever "bled through" from our multidimensional nature into ordinary consciousness would have to meet specific criteria. It couldn't be abstract or theoretical. It had to be demonstrable. Measurable. Universal.

First criterion: It must be at least partially observable, meaning we can sense at least part of it, which eventually morphed into: *It must cause observable physiological changes in the body.*

If this phenomenon was truly a bridge between dimensions, a way our infinite nature touches our finite form, it would have to show up in our biology. Our bodies, after all, are the instruments through which we experience three-dimensional reality. Any genuine bleed-through from higher dimensions would potentially translate into physical response.

Second criterion: Could it be some of what I was experiencing in deep meditative states? This wasn't theoretical. I had direct experience to work from. Ultimately, this is where my hypothesis began to take form.

If I was bypassing or 'unhooking' from the limitations of my default mind in meditation, what showed up next should be nothing at all. Yet there was only a void for a short while. Sometimes light came next. Then bliss, peace, and joy all descended, as well as an experience of oneness, abundance, and love. (Of course not in every meditation, but it certainly was repeatable if I went in deep.) How could I feel so full of something I didn't experience in my waking state?

In deep meditation, when ordinary awareness dissolved into something vaster, I encountered love, not as emotion, not as sentiment, but as a presence — a frequency that seemed to emanate from everything and nothing simultaneously. It was unconditional, impersonal, and absolutely undeniable.

At first, I wondered if this was simply a brain region activating and generating pleasant feelings, perhaps meditation triggering oxytocin or other neurochemicals that create sensations we interpret as love. But the quality of what I experienced felt fundamentally different from emotion. It wasn't my love for something. It was love as the substance of consciousness itself, perceived when my ordinary awareness became still enough to detect it. It was a bliss state for *All That Is*, that appeared every time I disconnected from the senses, leaving the awareness of the body-self behind.

Research confirms that meditation does increase oxytocin, the so-called 'love hormone', as well as dopamine and serotonin, all associated with feelings of pleasure, bonding, and connection.[1] Oxytocin itself is produced in the hypothalamus, a small but crucial region deep in the brain that regulates hormones and emotional responses.

Studies using functional magnetic resonance imaging (fMRI) show that meditation activates other specific brain regions, particularly the anterior cingulate cortex and dorsolateral prefrontal cortex -

areas associated with attention, emotional regulation, and self-awareness. [2]

The exact neural pathways connecting these phenomena in meditation aren't yet fully mapped. What's clear was that my practice of meditation, the state of focused attention and present-moment awareness, exactly when we unhook from the mind-body state, consistently correlates with elevated levels of neurochemicals associated with bonding, connection, and positive emotion.

This raised a compelling question: Was the neurochemical release *causing* the experience of love, even though it was produced in a different part of the brain than what is known to be activated in meditation? Or was perceiving the frequency of love instead *triggering* the neurochemical response? Perhaps the brain's production of oxytocin during meditation is the body's way of responding to — and translating — a frequency that exists beyond the physical.

The distinction matters. If meditation simply produces feel-good chemicals, then love is a biochemical event. But if meditation allows us to perceive an actual frequency, with neurochemicals being our body's response to that perception, then love is something we can learn to detect, like training our eyes to see in low light. I began to suspect the latter.

This experience kept pointing to the same possibility: what if love wasn't just a feeling meditation produced, but a frequency meditation allowed me to perceive? What if, in those expanded states, stripped away from the ego's limited focus and instead immersed in the conditions of a quiet, non-thinking, non-processing brain, I was detecting something that was always present but normally drowned out by the noise of ordinary conscious existence?

If love was truly what bled through from our multidimensional nature, it should become more apparent in deep meditative states, not less. And it did. Consistently. Undeniably. Powerfully.

Third criterion: It must be universally experienced across all cultures, times, and contexts. If this was truly fundamental to human nature, a manifestation of what we are beyond our limited perception, then every human being, everywhere, throughout all of history, would recognize it. And they have....pursued it, organized their lives around it. It couldn't be culturally specific or time-bound. It had to be essential.

Fourth criterion: Individual responses vary based on personal programming. I wasn't initially looking for this criterion but when I considered love as a candidate for what bleeds through, I realized it fit perfectly.

If love is a frequency we perceive rather than simply generate, then our response to perceiving it would depend entirely on our conditioning. It's like perfect pitch — the note is objectively what it is, but whether you can accurately hear it depends on your training and capacity.

Someone raised in an environment of safe, consistent love would have neural pathways already attuned to that frequency. When they encountered it, in meditation, in relationship, in nature, they'd recognize it easily, open to it naturally.

But someone whose early experiences taught them that "love" came with conditions, pain, or abandonment? Their nervous system might respond to genuine love with confusion, resistance, or even terror, not because the frequency itself changed, but because their receiver was calibrated differently. Same signal. Different receivers. Different responses.

This explained something I'd observed for years, which was why the same spiritual experiences that opened some people completely caused others to contract in fear, why love that healed one person triggered another, why meditation that revealed love to me might produce completely different experiences for someone else.

If love was truly a frequency bleeding through from our multidimensional nature, then of course responses to it would vary. We all carry different filters, different trauma patterns, different programming about whether it's safe to perceive love at all.

Whatever bled through from our multidimensional consciousness meet all four criteria and made love a perfect candidate to explore from all angles.

The Recognition

After exploring countless possibilities including intuition, consciousness itself, energy, presence, I kept returning to one answer that satisfied every requirement with startling precision—Love.

I'm not speaking of the romanticized, greeting-card version of love, and not the conditional love we so often mistake for the real thing. I was seeing love as a frequency, love as the vibrational signature of Source consciousness that our human awareness can actually detect. The moment this recognition landed, everything shifted.

Love causes measurable physiological changes. This isn't poetic metaphor; it's documented neuroscience. When we experience love, our bodies respond. Oxytocin floods our system, heart rate variability increases, stress hormones decrease, immune function improves. The physical body recognizes love and responds at the cellular level.

Love is perceivable in deep meditative states. In fact, it becomes overwhelming. Meditators across traditions - Buddhist, Hindu, Christian contemplative, Sufi - report the same experience. When awareness expands beyond the ordinary mind, what rushes in is love. Unconditional. Vast. Undifferentiated. Not love for something, but love as the fundamental quality of consciousness itself.

Love is universally sought. Every human culture, every spiritual tradition, every philosophical system recognizes love as primary. We structure our lives around it — who we love, how we express it, what we'll sacrifice for it. We seek it relentlessly, even when we've been hurt by its absence or its counterfeits. This seeking isn't learned behavior. It's recognition, the way a migrating bird recognizes magnetic north.

Responses to love vary wildly based on programming. Here's where it gets fascinating. The same frequency of love, received by different individuals, produces radically different reactions. Someone whose early life was filled with safe, consistent love might open to it effortlessly. Someone whose "love" came with conditions, violence, or abandonment might experience terror when genuine love approaches. Same frequency. Different receivers. Different responses. Love wasn't just one answer among many. It was the only answer that satisfied so many criterion.

A Case of The Broken Radio

Consider the case of people who have avoidant attachment — a pattern of behavior in relationships that stems from childhood trauma when a child doesn't receive consistent bonding from caregivers. Research shows these individuals develop altered oxytocin receptor systems and lower baseline oxytocin production throughout life.

If love were purely a neurochemical phenomenon and simply the result of oxytocin and dopamine creating pleasant sensations, we'd expect avoidant individuals to be relatively indifferent to love. Without the chemical reward system functioning properly, why would they seek something that doesn't produce the biochemical payoff? But that's not what we observe.

Avoidant individuals don't lack interest in love; they're deeply *conflicted* about it. They long for connection while simultaneously pushing it away. They recognize love when it's present, even as they struggle to trust it. They feel the profound absence of love, the wound of not having received it. This paradox reveals something crucial; they're detecting something their damaged biochemical systems can't fully process.

Think of it like a radio with a working receiver but a broken amplifier. The signal is there and they can sense its presence, but their internal equipment can't translate it into clear sound. They know something important is broadcasting and they just can't amplify it into the full experience their biology was designed to provide.

If love were only oxytocin, avoidant individuals would be like people born without taste buds — not conflicted about food, just indifferent. No reward, no seeking behavior. Yet avoidant attachment creates conflict, not indifference. They're responding to a frequency their perception can detect but their biochemistry can't properly amplify. The receiver works and they sense love's presence, but the translation system is impaired.

This suggests love exists as a perceivable frequency prior to and independent of our neurochemical response to it. The oxytocin, the dopamine, the serotonin are our body's way of amplifying and translating a signal that exists whether or not our biochemistry is functioning optimally.

Secure attachment doesn't create the capacity to detect love; it creates the capacity to amplify what we're already detecting. Avoidant attachment doesn't eliminate love perception but it damages the amplification system.

The frequency is always broadcasting. The question is whether our receivers are tuned to it and whether our amplifiers can translate it into the full-body experience of connection, safety, and bonding.

This explains something meditation teachers have known for millennia. Even people with severe attachment trauma can, through practice, learn to perceive and rest in love as a fundamental quality of consciousness. They're not creating the love through meditation but instead they're tuning their perception to detect what was always present, regardless of whether their oxytocin systems learned to respond properly in childhood. The practice doesn't fix the amplifier (though that may gradually heal through neuroplasticity). The practice trains the receiver to detect the signal so clearly that even a broken amplifier can't completely obscure it.

Some might argue that avoidant individuals seek love not because they perceive a frequency, but because they've learned cognitively that love is valuable. This doesn't account for the compulsive quality of their seeking, the way they pursue connection despite it terrifying them, despite their biochemistry providing minimal reward. This isn't rational cost-benefit analysis. This is perception (I detect something real and compelling) conflicting with protection (my system can't safely process it). The very irrationality of the pattern of seeking what you simultaneously sabotage, reveals something beyond thought or chemistry. It is the detection of a signal their damaged equipment struggles to translate.

What This Means

If love is what bleeds through from our multidimensional nature into our limited 3D awareness, then love isn't primarily an emotion. It's a sense. Our true sixth sense.

Just as our eyes detect light and our ears detect sound, we have the capacity to detect love — the vibrational frequency of Source consciousness as it expresses through multiple dimensions and penetrates into our ordinary awareness.

Most of us have never been taught to recognize love as perception. Of course not! Why would we? We think of it as feeling, as relationship, as something that happens to us rather than something we can consciously attune to. We're like people born with functioning eyes who've never been told they can see, and mistaking the blurry shapes and colors for something other than visual information.

But once you recognize love as a sense, as a capacity for perception that can be developed, refined, and mastered, everything changes. You stop waiting to fall into love and start learning to tune into love. You stop looking for love from others and start detecting love as the fundamental frequency underlying all of reality. You stop thinking you've lost love when a relationship ends and start recognizing that love, the frequency, the sense, the bleed-through from your multidimensional nature - is always present, always broadcasting, always available to those who learn to perceive it.

This is what I'd been searching for. Not a philosophy. Not a practice. But the recognition of what we already are, multidimensional beings whose truest nature seeps into our limited awareness as love, detectable by those who learn to sense it.

It's ok if you're not convinced yet. I wasn't either. We're moving forward to explore the science soon, after we lay a few more foundations.

CHAPTER THREE
Understanding Dimensional Intersections

If love is the trace of something higher-dimensional passing through us, then we should be able to find clues — patterns, consistencies, signatures — that point toward its true nature. Just as the U-shaped cutout on the plexiglass plane revealed something about the rotating rod, the way love shows up in our lives may reveal something about the structure of consciousness itself.

To understand the clues, we first have to ask, what are we looking at? If we accept the premise that higher dimensions exist, and modern physics increasingly suggests they do, then we must also accept that these dimensions are not separate from us. They are not "somewhere else out there." They are here, layered into the same spacetime we inhabit, but vibrating at frequencies our biology is not optimized to detect.

Think of it as a radio, which doesn't create the music. It tunes into frequencies already present in the electromagnetic field. Change the dial, and you access a different broadcast, not because the station suddenly appeared, but because you shifted your receiver into alignment with it.

Does consciousness work the same way? What if love is one of the frequencies we can tune into, not something we generate, but something we perceive when our internal receiver is calibrated correctly? This chapter explores the evidence for that possibility. We'll look at how love behaves, where it shows up, and what patterns emerge when we stop thinking of it as a psychological construct and start treating it as a dimensional phenomenon.

The Consistency of the Trace

One of the most compelling pieces of evidence that love is more than a neurochemical accident is its consistency across cultures, across time, and across states of consciousness.

Every human culture has words for love. Every spiritual tradition points toward it as the ultimate truth. From the Sufi poets to the Buddhist teachings on metta (loving-kindness), from the Christian mystics to the Advaita sages, the message is the same. Love is not just a feeling. It is the ground of being itself.

The love they're describing is not the same love that shows up in a neurochemistry textbook. Dopamine, oxytocin, and vasopressin can explain pair bonding. They can explain attachment between mother and child. They can even explain the euphoria of new romance. But they cannot explain the love that arises in deep spiritual practice and meditation when there is no object, no other person, no biological imperative. They cannot explain the love that floods through a Reiki session, or the palpable warmth that radiates from an enlightened being who has transcended all personal attachment.

This is not the love of survival and not the love of reproduction. This is something else entirely. Yet it feels the same in the body, or rather, it feels more real than romantic love, more stable than attachment, more foundational than desire. It is love without an object or a reason. It's love that simply is.

What if this is the actual frequency of love, and what we call "romantic love" or "familial love" is simply a trace of that frequency filtered through the lens of biology, psychology, and personal history? What if, in other words, the love we seek is not the thing itself, but one of the shadows it casts when it intersects our limited perceptual field?

The Rod Rotates, the Plane Perceives

Let's return to the metaphor. The rod is always rotating. It doesn't start and stop. It's in constant motion above the plane, and as it rotates, different parts of it intersect the plexiglass at different angles. Sometimes the intersection is shallow, just a faint curve. Sometimes it's deep a bold, unmistakable U-shape. But the rod itself never changes. Only the angle of intersection changes. Now translate this to love.

The field of love, the higher-dimensional reality, the unified field, the quantum foam, the Source frequency, whatever name you want to give it, is always present. It doesn't come and go. It doesn't depend on whether you're in a relationship, whether you're meditating, whether you're happy or sad. It's the substrate. The ground. The fabric of reality itself. But our perception of it changes based on the state of our nervous system, the coherence of our biofield, the clarity of our awareness, and the degree to which we are aligned with that frequency.

When we say "I feel love," what we're really describing is a moment when the perception of intersection became optimal. The trace became undeniable. But the rod was always there.

When we say "I fell out of love," or "the love faded," what we're describing is not the disappearance of love itself, but a shift in the angle of our perception. We moved. The plane tilted. The intersection became shallow again, and the trace became harder to detect. This is why love can feel both constant and fleeting. The field is constant. Our access to it is not.

Where the Trace Shows Up

If this hypothesis is correct and if love is a higher-dimensional field intersecting our biology, then we should be able to identify where and how the trace appears. And indeed, there are patterns.

1. In Deep Meditation

One of the most common reports from experienced meditators is the arising of unconditional love. This is not love for something, but love as a state. It's a boundless, objectless warmth that fills the chest, expands through the body, and seems to radiate outward into the space around them.

Neuroimaging studies of meditating monks show increased activity in the prefrontal cortex and anterior cingulate, regions associated with attention, compassion, and self-regulation. But they also show something else which is a coherence in brainwave patterns. The brain begins to oscillate in sync, producing high-amplitude gamma waves (30-100 Hz) that are rarely seen in ordinary waking consciousness.

This coherence is not just happening in the brain. Studies have shown that the heart's electromagnetic field also becomes more coherent during states of love and compassion as smooth, sine-wave-like rhythms replace the jagged, chaotic patterns of stress or distraction.

Maybe this coherence is not the cause of love, but the condition that allows love to be sensed more clearly. What if, when the nervous system becomes still and the biofield becomes coherent, we are simply tuning in to a frequency that was always there?

2. *In the Presence of Certain Beings*

Another clue is the phenomenon of darshan — the experience of being in the presence of a saint, guru, or enlightened being and feeling an overwhelming wave of love, peace, or bliss.

This is not mere projection. It's a felt experience reported by thousands of people across cultures and spiritual traditions, including myself. Whether it's Amma the "Hugging Saint," Ramana Maharshi, or Thich Nhat Hanh, people describe the same thing — a palpable field of love radiating from these beings that can be felt even from a distance.

What's fascinating is that this experience doesn't require belief. Skeptics feel it. Children feel it. Animals feel it. It's not a matter of faith; it's a matter of resonance.

If love is a dimensional field, then perhaps these beings are not "more loving" in the emotional sense, but rather more coherent. Their nervous systems, their biofields, their entire being is aligned with Source frequency to such a degree that they transmit its essence of love, the way a tuning fork resonates and causes nearby forks to vibrate in sympathy.

3. *In Energy Healing*

Reiki practitioners, energy healers, and bodyworkers consistently report feeling love — not as an emotion they're trying to cultivate, but as a presence that arises spontaneously during the work. Recipients of the work report the same thing: a sensation of warmth, safety, and often profound love, even when the practitioner is a stranger.

This is not touch therapy or massage. The hands are often hovering above the body, not making contact. And yet, something is transmitted, something measurable, even.

Studies have detected pulsing magnetic fields emanating from the hands of healers during sessions — fields that are significantly stronger than the baseline magnetic field of the human body. These fields oscillate at frequencies that correspond to the body's natural healing rhythms (0.3-30 Hz).

What if these healers are not "generating" love, but channeling it? They could be simply attuned to a frequency that, when transmitted, allows the recipient's nervous system to also tune in.

4. *In Moments of Grace*
Finally there are the spontaneous moments — the ones that can't be explained by meditation practice, spiritual lineage, or intentional healing. These are moments when love just floods in for no apparent reason. They include walking in nature, holding a newborn, watching the sunrise. Sometimes even when losing a loved one and, in the midst of grief, people report feeling held by something vast and benevolent.

These moments are often described as grace, an unearned, unexpected gift. They're not gifts at all. What if they're simply moments when the angle of intersection became optimal — when the noise of the mind quieted, when the nervous system relaxed, when the heart opened — and suddenly, the trace became visible? What if grace is just clarity, resonance with our true Source-infused nature?

The Pattern Emerges

When we look at all these clues together, a pattern begins to emerge. Love arises most powerfully in states of coherence when the brain, heart, and nervous system are aligned and oscillating in sync.

Love is felt most clearly in the absence of mental noise when the thinking mind quiets and attention becomes spacious.

Love is transmitted between beings, not through words or touch, but through resonance, through the biofield, through some form of energetic entrainment.

Love is objectless at its purest, not love for someone or something, but love as a background presence, a field that is always there when we are still enough to notice it.

This is not the behavior of a neurochemical reaction. This is not the behavior of a psychological construct. This is the behavior of a frequency — a vibrational field that we can tune into when the conditions are right. And if love is a frequency, then it makes sense that we would need a way to detect it, not just intellectually, but somatically, and not just conceptually, but experientially. We would need a sense.

The Question Deepens

So the question becomes, what part of us is doing the sensing? It's clearly not the eyes; we don't see love. It's not the ears; we don't hear it. It's not the tongue, the nose, or the skin in the conventional sense. And yet, we know when it's there. We feel it in the body. We recognize it instantly, even when it arrives unexpectedly.

It changes our physiology including our heart rate, our breathing, our brainwaves. It shifts our nervous system from sympathetic (fight-or-flight) to parasympathetic (rest-and-digest). It releases neurochemicals. It activates specific brain regions. So something is detecting it. Something is transducing it — turning an external stimulus into an internal experience. But what? The answer, I believe, lies at the intersection of three emerging fields of science.

Biofield science — the study of the electromagnetic and subtle energy fields surrounding and permeating the body.

Quantum biology — the discovery that quantum coherence, entanglement, and tunneling play a role in biological processes like photosynthesis, bird navigation, and possibly human consciousness.

Interoception — the sense of the internal state of the body, mediated by specialized receptors that detect heart rate, breath, gut signals, and other visceral sensations.

It is possible that love is sensed through a combination of all three. Potentially, it's detected by the biofield, transduced through quantum processes in our cells, and felt through the interoceptive pathways that connect the body to the brain. What if, in other words, we already have a sixth sense and we've just been calling it by the wrong name?

The Cutout Points the Way

The U-shaped trace on the plexiglass plane cannot reveal the full shape of the rod. But it can point us in the right direction. It can tell us that something is there, even if we can't see it fully. It can give us clues about its motion, its angle, its presence.

And so it is with love. The longing we feel, the moments of connection, the ache of separation, the peace of meditation, the warmth of a healing session are all traces. They are the U-shaped cutouts left behind as something vast, something higher-dimensional, something real passes through us. We can study the traces. We can map them. We can learn to recognize them more clearly. And in doing so, we can begin to understand not just what love is, but how we are designed to perceive it.

In the next chapter, we will explore the mechanics of sensing and how the body turns stimuli into experience, and why love may fit the criteria of a sense better than we ever imagined.

But for now, simply consider this: if the trace is real, then so is the rod. And if the rod is real, then we are not alone in a cold, indifferent universe. We are immersed in love. We just haven't learned how to see it yet.

CHAPTER FOUR
Introducing the Non-Mystical Sixth Sense

We are taught, from childhood, that we have five major senses — sight, hearing, touch, taste, and smell. These are the tools through which we navigate the world, interpret stimuli, and construct our understanding of reality. They are so fundamental to our experience that we rarely question them. We simply use them, the way we breathe, automatically, unconsciously, without wondering how or why they work.

We've been missing one. I propose there is a sixth sense, not in the mystical, paranormal way the phrase is often used, but in the literal, biological sense — a faculty by which the body detects and responds to an external stimulus, processes it through neural pathways, and translates it into a felt experience. This is a testable hypothesis. And to test it, we need to begin with a clear definition, so what exactly, is a sense?

The Anatomy of a Sense

According to biology, a sense is *a faculty by which the body perceives an external stimulus.* To understand what that means, we need to break it down into its component parts. A true sensory system must meet the following criteria:

1. *There must be a stimulus.* Something in the external environment — light waves, sound waves, chemical molecules, pressure, temperature, that can be detected.

2. *There must be a receptor.* Specialized cells or proteins that respond to that specific type of stimulus. For example, *p*hotoreceptors in the retina detect light. Hair cells in the cochlea detect sound vibrations. Chemoreceptors in the nose and tongue detect molecules. Mechanoreceptors in the skin detect pressure and touch.

3. *There must be a transduction process.* The receptor must convert the external stimulus into an electrical signal that the nervous system can understand. This is the moment when physical energy becomes biological information.

4. *There must be a neural pathway.* The electrical signal must travel from the receptor to the brain, often through multiple synaptic relays, where it is processed, interpreted, and integrated with other sensory data.

5. There must be a conscious perception. The final result is a felt experience — a sensation, a qualia, a knowing that something has been detected. You see the light. You hear the sound. You feel the touch.

If love is a sense, then it must meet all five criteria. So let's go through them, one by one.

Criterion 1: Is There a Stimulus?

The first question is, what is the external stimulus that triggers the sensation of love?

This is where conventional science stumbles because if you ask a neuroscientist what causes love, they will tell you dopamine, oxytocin, vasopressin. They will point to the limbic system, the ventral tegmental area, the nucleus accumbens. They will describe love as a response, a reaction to seeing a beautiful face, holding a baby, or receiving a kind touch.

But this is circular reasoning. It's like saying, "We see because photons hit the retina, which triggers a neural response." Yes, that's true, and the photons are the stimulus. The neural response is the transduction. The two are not the same.

So what is the stimulus for love? The stimulus is not a person, not a touch, not a memory. The stimulus is a vibrational frequency — an energetic signature present in the environment, detectable through the biofield, and corresponding to what we might call coherence or resonance with Source. In these tender moments, we leave behind our linear thinking, our reasoning, and our interpretive subconscious, and in flows the true nature of our beingness, that which is closely aligned with Source. This is exactly what happens in meditation. This may sound abstract if you're not regularly experiencing this as a meditator, but there is evidence for it.

Studies have shown that the heart generates an electromagnetic field that extends up to 3 feet beyond the body and can be detected by magnetometers. This field changes its pattern based on emotional state — becoming smooth and coherent during states of love, gratitude, and compassion, and jagged and chaotic during stress or anger.

Biofield effects are measurable. Healers working with Reiki or Therapeutic Touch produce pulsing magnetic fields from their hands at frequencies that correspond to the body's natural healing rhythms (0.3-30 Hz). These fields can be detected even when the hands are not touching the body.

Quantum coherence occurs in biological systems. Studies in quantum biology have shown that photosynthesis, bird navigation, and even enzyme activity involve quantum processes — coherent oscillations at the cellular level that allow for non-local information transfer.

What if love, too, is a form of quantum coherence? What if it is a resonant frequency in the unified field — the zero-point field, the quantum vacuum, the fabric of spacetime itself (all that I have simply termed Source for this book) — and what we experience as love is simply our body-mind detecting and aligning with that frequency?

If this is true, then the stimulus is not a person. The stimulus is a field state. And certain conditions such as proximity to a coherent being, deep meditation, energy healing, moments of grace all allow us to tune into that state more clearly.

Criterion 2: Is There a Receptor?

If love is a stimulus, then there must be something in the body designed to detect it. But what? This is where things get interesting because there may be multiple types of receptors involved, working at different levels of the system.

1. The D3 Dopamine Receptor

Dopamine is often called the "love drug," but that's a misnomer. Dopamine is a neurotransmitter — a chemical messenger that passes signals from one neuron to another. It's involved in movement, motivation, pleasure, and reward. Dopamine itself is not love.

However, there is a specific subtype of dopamine receptor — the D3 receptor (Drd3) — that is most strongly expressed in the limbic system, particularly in two regions called the islands of Calleja and the nucleus accumbens. These regions are involved in the reinforcing effects of reward-type activities and pleasure.

Interestingly, the D3 receptor is different from the other dopamine receptors. It has a unique structure, a unique distribution, and a unique function. It is not just about pleasure — it is about deep, sustained states of reward and well-being. It is about the feeling of rightness, of homecoming, of ‘this is what I've been searching for’.

Could the D3 receptor have evolved specifically to detect the love frequency? What if it is the biological interface between the vibrational field and the nervous system?

2. Interoceptive Receptors Interoception is the sense of the internal state of the body. It's how you know you're hungry, thirsty, tired, or in pain. It's mediated by specialized receptors throughout the body, in the heart, the gut, the lungs, the blood vessels, that detect things like heart rate, blood pressure, gut motility, and breath rhythm.

Yet interoception is not just about survival. It's also deeply connected to emotion. Studies have shown that people with greater interoceptive awareness, those who can accurately sense their own heartbeat, for example, also report more intense emotional experiences, both positive and negative.

If love is a tenet of Source, it would not be a far stretch to identify love as a form of refined interoception — not just sensing the internal state of your own body, but sensing the resonance between your body and the field? Perhaps when you *feel* love, what you're actually detecting is the coherence of your biofield aligning with something larger.

3. Biofield Receptors (Cryptochromes and Magnetoreception) This is the most speculative and also the most exciting possibility.

There is evidence that humans, like birds, may have proteins called cryptochromes in our retinas that are sensitive to magnetic fields. These proteins are thought to play a role in circadian rhythms, but they may also function as magnetoreceptors, allowing us to detect subtle variations in the Earth's magnetic field or the biofields of other living beings.

If this is true, we may literally be able to sense the electromagnetic field of another person — and that field, when coherent, may register as love. This would explain why we can "feel" when someone is staring at us. We can sense the presence of others even when we can't see or hear them. Certain spaces feel "sacred" or "charged" with energy. We feel love more strongly in the presence of beings whose biofield is highly coherent (enlightened beings, healers, etc.).

Criterion 3: Is There a Transduction Process?

Yes! And this is where the biochemistry comes in. When the stimulus (the love frequency) is detected by the receptors (D3, interoceptive, biofield), it triggers a cascade of biochemical events.

Dopamine is released in the nucleus accumbens and islands of Calleja, binding to D3 receptors and creating the sensation of bliss, reward, and deep satisfaction. Oxytocin and vasopressin are released in the hypothalamus, promoting bonding, trust, and social connection.

The heart's electromagnetic field becomes coherent, shifting from chaotic to smooth sine-wave patterns. Brainwaves synchronize, particularly in the gamma range (30-100 Hz), indicating high-level integration across brain regions. The parasympathetic nervous system activates, shifting the body into a state of rest, healing, and receptivity.

This is the transduction — the moment when the external stimulus (the love frequency) is converted into an internal, biological response. This would mean that the biochemistry is not the cause of love. It is the body's way of registering and responding to the stimulus. Just as photons hitting the retina trigger a biochemical cascade that results in vision, the love frequency hitting the biofield triggers a biochemical cascade that results in the sensation of love.

Criterion 4: Is There a Neural Pathway?

Absolutely. Neuroimaging studies have shown that when people experience love, whether romantic, maternal, or unconditional, specific brain regions light up. These are the ventral tegmental area (VTA), which produces dopamine, the nucleus accumbens, which processes reward, the anterior cingulate cortex, which regulates emotion and social connection, the prefrontal cortex, which integrates complex information and makes meaning, and finally the insula, which processes interoceptive signals and bodily awareness.

These regions are not isolated. They form a network, a neural circuit that carries the signal from the initial detection (in the biofield, in the heart, in the gut) to the final conscious perception (in the cortex).

This is the pathway. It is as real and measurable — as real as the pathway that carries visual information from the retina to the occipital lobe.

Criterion 5: Is There a Conscious Perception?

Yes, and this is the most undeniable part of all. We know when love is present. We feel it. It has a quality, a texture, a presence that is unmistakable. It is not the same as happiness, or excitement, or attraction. It is its own distinct experience, a qualia that cannot be reduced to any other sensation.

The remarkable part that goes against what science says, is that we recognize it even when it has no object. In deep meditation, there is no person to love, no situation to be grateful for, no memory to cherish. There is just love, arising spontaneously, filling the body, radiating outward. This is not an emotion triggered by a thought. This is not a feeling generated by the mind. This is a sensation — a direct, unmediated perception of something present in the field. This is what distinguishes love as a sense. Just as you can see light without thinking about light, you can sense love without thinking about love. It simply is.

The Checklist

Let's review. For love to qualify as a sense, it must meet five criteria:

1. Stimulus Yes — a vibrational frequency/field state

2. Receptor Yes — D3 dopamine receptors, interoceptive receptors, possible magnetoreceptors.
3. Transduction Yes — biochemical cascade (dopamine, oxytocin, heart coherence, brainwave synchrony)
4. Neural Pathway Yes — VTA → nucleus accumbens → anterior cingulate → prefrontal cortex → insula
5. Conscious Perception Yes — the unmistakable, objectless sensation of love

Love meets all five criteria. It is not an emotion. It is not a feeling about something. It is a sense — a perceptual faculty by which the body detects and responds to an external vibrational stimulus. And if this is true, then everything changes.

The Implications

If love is a sense, then we are not generating it. We are detecting it. The love is already there, in the field, and we are simply tuning into it when the conditions are right.

We can train it. Just as you can improve your eyesight with practice, or refine your sense of taste through attention, you can sharpen your ability to sense love. Meditation, coherence practices, energy work are not ways to *create* more love. They are ways to clear the signal.

We are not alone. If love is a field, a dimensional frequency, a quantum coherence, a resonance with Source, then we are immersed in it at all times. Separation is an illusion created by the limits of our perception. The longing we feel is not a sign that love is absent. It is a sign that we are sensing it, but not fully.

The search makes sense. Why do humans chase love so relentlessly? Because we are biologically, neurologically, and energetically designed to. The D3 receptors, the interoceptive pathways, the biofield sensitivity all point towards a system optimized to detect and align with the love frequency. We are not broken for wanting it. We are designed for it.

Most profoundly, love is not a reward for being good, or lucky, or worthy. It is a tenet of reality — as fundamental as light, as pervasive as gravity, as real as the air we breathe. It was never something we had to earn. It was always something we had to remember how to sense.

The Forgotten Sense

For too long, we have treated love as a fraction of its true essence, as the part that was visible to us in our Flatland world. It's completely understandable. But pieces don't always tell the whole story; they only provide clues. Love seemed to be tied to our emotions, as something fleeting, subjective, unreliable. We have reduced it to biochemistry, as a trick of evolution to keep us bonding, reproducing, surviving. We have made it contingent, that it is something we receive only if we are lovable, only if we find the right person, only if we do everything right.

But what if we've had it backwards this entire time?

What if love is not something we create, but something we perceive? What if the reason we search for it so desperately is not because we lack it, but because we sense it, like a distant frequency just beyond the edge of hearing, like a light at the edge of vision, like a scent we can't quite place but know we've smelled before?

Perhaps the reason every spiritual tradition points toward love as the ultimate truth is because they are all describing the same sense, the one we forgot we had, the one that, when activated, reveals the true structure of reality.

This is the forgotten sense.

In the chapters to come, we will explore how to reclaim it, how to understand its biology, its physics, its practice. We will look at how the body senses love, how the heart transmits it, how plants and animals respond to it, and how consciousness itself may be the field in which love arises.

But for now, simply *sit with this possibility:* Love is not what you think. It is not what you feel. It is what you sense — a dimension of reality you were always designed to perceive. And once you remember how to sense it, everything changes.

CHAPTER FIVE
Sensory Science and the Body-Mind

To understand how love might function as a sense, we first need to understand how our known senses work. What, exactly, happens in the body when we see, hear, touch, taste, or smell? And more importantly, what are the patterns? What do all sensory systems have in common, and how might love fit into that framework?

This chapter is a tour through the mechanics of perception, not to drown in technical detail, but to reveal the elegance of how the body translates the external world into internal experience. Because once we see the pattern, we can recognize it when it shows up in unexpected places. Like love.

The Universal Pattern:

Stimulus → Receptor → Transduction → Pathway → Perception

Every sense follows the same basic architecture:

A stimulus exists in the environment — light waves, sound waves, molecules, pressure, temperature.

A receptor detects it — specialized cells or proteins that respond only to that specific type of stimulus.

Transduction occurs — the physical stimulus is converted into an electrical signal the nervous system can understand.

The signal travels along a neural pathway — from the receptor to the brain, often through multiple relay stations.

Perception arises — the conscious experience of seeing, hearing, feeling, tasting, or smelling.

This is not random or accidental. This is the body's way of knowing the world by turning energy into information, information into experience, experience into meaning. Let's walk through each of the five senses and see how this pattern plays out.

Sight: Turning Light Into Vision

The Stimulus

Light — electromagnetic radiation traveling at 299,792 kilometers per second, oscillating at different frequencies that we perceive as different colors.

The important thing to remember is that we only see a tiny sliver of the electromagnetic spectrum. Visible light ranges from about 380 nanometers (violet) to 700 nanometers (red). Beyond that, we can't see infrared, ultraviolet, X-rays, radio waves, or gamma rays. They exist and are very real but our biology is not tuned to detect them.

Already, this tells us our perception is always limited. We don't see reality — we see the part of reality our receptors are designed to detect.

The Receptor

In the retina at the back of the eye, there are two types of photoreceptors.

Rods — sensitive to dim light, responsible for night vision and peripheral awareness. About 120 million of them.

Cones — sensitive to color, concentrated in the center of the retina (the fovea). About 6-7 million of them, divided into three types: red, green, and blue.

Each photoreceptor contains a protein called rhodopsin (in rods) or photopsin (in cones), which changes shape when struck by a photon of light.

The Transduction

When light hits rhodopsin, the protein undergoes a structural change — it literally bends. This triggers a cascade of chemical reactions that ultimately cause the photoreceptor to stop releasing neurotransmitter. (Yes, the signal is a decrease in activity, not an increase. The body is clever.)

This change is detected by bipolar cells, which pass the signal to ganglion cells, whose axons form the optic nerve.

The Pathway

The optic nerve carries the signal from the retina to the thalamus (a relay station in the center of the brain), and from there to the visual cortex at the back of the brain (the occipital lobe).

The signal doesn't stop there. The visual cortex has multiple layers, each processing different aspects of the image — edges, colors, motion, depth. The brain is not passively receiving an image; it is actively constructing one, integrating data from millions of photoreceptors into a coherent picture of the world.

The Perception

You see. You perceive light, color, form, movement. Yet what you are actually experiencing is not the light itself — it's the brain's interpretation of electrical signals generated by photoreceptors responding to light. Vision is a sensory hallucination. A very accurate one, but a hallucination nonetheless.

Hearing: Turning Vibration Into Sound

The Stimulus

Sound waves are pressure waves traveling through air (or water, or solid objects) at different frequencies. We hear frequencies roughly between 20 Hz and 20,000 Hz. Below that is infrasound (felt more than heard). Above that is ultrasound (used by bats and dolphins, but inaudible to us). Again, we only detect a narrow band of what exists.

The Receptor

Sound waves travel through the ear canal and vibrate the eardrum, which moves three tiny bones (the malleus, incus, and stapes — the smallest bones in the body). These bones amplify the vibration and transmit it to the cochlea, a snail-shaped fluid-filled structure in the inner ear.

Inside the cochlea, there is a membrane called the basilar membrane, covered with thousands of tiny hair cells. These are the actual receptors. Each hair cell is tuned to a specific frequency, arranged in order from high to low along the length of the cochlea (this is called tonotopic organization).

When the fluid in the cochlea vibrates, it causes the hair cells to bend. And when they bend, they open ion channels, triggering an electrical signal.

The Transduction

The bending of the hair cell is the moment of transduction, and the mechanical movement becomes electrical signal.

The Pathway

The electrical signal travels along the auditory nerve to the brainstem, then to the thalamus, and finally to the auditory cortex in the temporal lobe.

The Perception

You hear. You perceive pitch, volume, timbre, location. But again, what you are experiencing is not the sound itself — it's the brain's reconstruction of the pattern of electrical signals generated by vibrating hair cells. Hearing is also a sensory hallucination.

Touch: Turning Pressure Into Sensation

The Stimulus

Pressure, vibration, stretch, texture — any mechanical force applied to the skin.

The Receptor

The skin is covered with mechanoreceptors which are specialized nerve endings that respond to different types of mechanical stimulation.

Merkel cells — detect fine details and textures (like reading Braille).

Meissner's corpuscles — detect light touch and changes in texture (like stroking a soft fabric).

Pacinian corpuscles — detect deep pressure and high-frequency vibration (like the buzz of a phone in your pocket).

Ruffini endings — detect skin stretch and sustained pressure (like gripping a handle).

Each type of receptor has a slightly different structure, optimized for its specific function.

The Transduction

When pressure is applied to the skin, the mechanoreceptor deforms, opening ion channels and generating an electrical signal.

The Pathway

The signal travels along sensory neurons to the spinal cord, then up to the thalamus, and finally to the somatosensory cortex, a strip of brain tissue that maps the entire body surface. (Fun fact: The lips and hands take up a disproportionately large area of this map, because they are so sensitive.)

The Perception

You feel the touch. You perceive pressure, texture, temperature, pain. But once again, what you are experiencing is not the touch itself — it's the brain's interpretation of electrical signals generated by mechanoreceptors responding to pressure. Touch is also a sensory hallucination.

Taste and Smell: Turning Molecules Into Flavor

These two senses work together so closely that they're often grouped as the chemical senses. Both involve detecting specific molecules in the environment and translating their chemical structure into neural signals.

Taste (Gustation)

The Stimulus: Tastants — chemical molecules dissolved in saliva.

The Receptor: Taste receptor cells, located in taste buds on the tongue, palate, and throat. There are five basic taste categories, each detected by a different type of receptor:

Sweet (sugars, some amino acids)
Salty (sodium chloride and other salts)
Sour (acids, hydrogen ions)
Bitter (alkaloids and toxins)
Umami (glutamate, the "savory" taste)

The Transduction: When a tastant molecule binds to its receptor, it triggers a cascade of chemical reactions inside the taste receptor cell, ultimately causing the release of neurotransmitter onto a sensory neuron.

The Pathway: The signal travels along cranial nerves to the brainstem, thalamus, and gustatory cortex (near the somatosensory cortex).

The Perception: You taste.

Smell (Olfaction)

The Stimulus: Odorants — volatile chemical molecules carried through the air.

The Receptor: Olfactory receptor neurons in the olfactory epithelium (a patch of tissue high up in the nasal cavity). Humans have about 400 different types of olfactory receptors, each tuned to a different molecular shape. Each odorant activates a unique pattern of receptors, creating a kind of molecular fingerprint.

The Transduction: When an odorant binds to its receptor, it triggers a G-protein cascade (similar to vision), generating an electrical signal in the olfactory neuron.

The Pathway: Olfactory neurons send their axons directly to the olfactory bulb (part of the brain), bypassing the thalamus. From there, signals go to the piriform cortex (primary smell center) and the amygdala and hippocampus (emotion and memory centers). This is why smell is so deeply tied to memory and emotion — it has a direct line to the limbic system.

The Perception: You smell. And often, you simultaneously remember — a scent can bring back a memory from childhood with startling vividness.

The Pattern Revealed

Do you see the pattern? Every sense starts with energy or molecules in the environment. Every sense has specialized cells designed to detect that specific stimulus. Every sense converts the stimulus into electrical signals. Every sense sends those signals to the brain for processing. And every sense results in a conscious experience — a qualia — that is fundamentally different from the raw data.

The world we experience is not the world that exists. It is a translation, a sensory hallucination constructed by the brain from patterns of electrical activity. *The brain doesn't know the difference between "real" and "hallucinated." It only knows electrical signals.* If the signals follow the right pattern, the brain will construct a perception, whether the stimulus came from outside the body or was generated internally.

This is why you can *see* things in dreams, *hear* a song in your head, or *feel* a phantom limb after amputation. The pattern is what matters, not the origin.

So Where Does Love Fit?

If love is a sense, it must follow this same pattern. So let's map it:

Sense = Stimulus → Receptor → Transduction Pathway → Perception

Stimulus: Vibrational frequency / biofield coherence

Receptor: D3 dopamine receptors, interoceptive receptors, possible magnetoreceptors

Transduction Pathway - Field resonance → biochemical cascade (dopamine, oxytocin, heart coherence) → electrical signal VTA → nucleus accumbens → anterior cingulate → prefrontal cortex → insula

Perception = Love

It all fits. Love has a stimulus (the waveform or expression of the Source field). It has receptors (D3, interoceptive pathways, biofield sensitivity). It has transduction (biochemical cascade). It has a pathway (limbic system → cortex). And it has perception (the unmistakable sensation of love).

The only difference is that we haven't been taught to recognize it as a sense. We've been taught to think of it as an emotion, a feeling, a psychological construct. But emotions come and go based on thoughts. Feelings are interpretations of bodily states. Love, at its core, is neither of these. It is a direct perception — a sensory event.

If love is a sense, then:

It is not a sign of weakness or neediness. It is a biological faculty, like vision or hearing. To long for love is not to be broken; it is to be sensing something real.

It is not subjective or arbitrary. Just as two people looking at a red apple both see red (assuming their photoreceptors work), two people in a coherent field can both sense love. It is intersubjective, shared, consistent, detectable.

It can be refined. Just as you can train your palate to detect subtle flavors in wine, or your ear to distinguish pitches in music, you can refine your ability to sense love. Meditation, coherence practices, energy work — these are not creating love. They are clearing the noise so the signal comes through more clearly.

It is always present. Light exists whether or not you open your eyes. Sound exists whether or not you listen. And love exists whether or not you are tuned to its frequency. The field is constant. Our access to it is not.

The Body-Mind as Instrument

The final, crucial insight is that your body is not just a vehicle for your consciousness. It is a sensory instrument — a finely tuned receiver designed to detect and transduce multiple layers of reality.

Your eyes detect light. Your ears detect sound. Your skin detects pressure. Your nose detects molecules. Your tongue detects chemicals. And your heart, your biofield, your limbic system, your entire nervous system are tuned to detect love — love as a vibrational frequency, a coherence in the field, a resonance with Source, a dimensional signal passing through the plane of your awareness, that gives you tangible clues to the nature of this reality construct, and the field from which everything is created.

And just like the Flatlanders watching the U-shaped cutout form and disappear, you are watching the trace of something vast, something real, something that was always there. You just didn't know you had the sense to perceive it until now.

CHAPTER SIX
The Biology of Love

If love is a sense, then it must leave a biological trail. And it does. When you fall in love, the kind of love that most of humanity recognizes, it is no different. Your body knows it before your mind does. Your heart races. Your palms sweat. Your pupils dilate. Your breathing quickens. You feel light, expansive, almost floating. Or perhaps you feel grounded, safe, held. Either way, something physiological is happening — something measurable, repeatable, undeniable.

For decades, scientists have been mapping this biology of love. They've identified the neurotransmitters, the brain regions, the hormones, the neural circuits. They've scanned the brains of people in love and watched them light up like fireworks. They've measured heart rates, cortisol levels, oxytocin surges. They've built a model, a biochemical explanation for why love feels the way it does.

The model is not wrong. It's potentially just incomplete. This chapter is a tour through the biology of love — not to dismiss any science, but to show where it points beyond itself. The chemicals are real. The neural pathways are real. The physiological changes are real. But I believe they are not the cause of love; they are the body's way of registering it.

The Usual Suspects: Dopamine, Oxytocin, and Vasopressin

Let's start with the three neurochemicals most commonly associated with love.

Dopamine: The Motivation Molecule

Dopamine is often called the "love drug" or the "reward chemical," but that's misleading. Dopamine doesn't create pleasure — it creates motivation. It's the neurotransmitter of wanting, not liking. It's what drives you to pursue a goal, seek a reward, anticipate something good.

When you see someone you're attracted to, dopamine floods the brain. When you think about them, dopamine spikes. When you're about to see them, dopamine surges. This is why new love feels so intoxicating. It's a dopamine high, similar to what happens with cocaine, gambling, or any other addictive stimulus.

The key is that dopamine is released in response to a stimulus — it's not the stimulus itself. Just as your pupils dilate in response to light (but are not the light), dopamine is released in response to something, and that something includes love.

So what triggers the dopamine? The conventional answer is seeing an attractive face, receiving a text message, anticipating a date. But that doesn't explain unconditional love. It doesn't explain the dopamine surge that happens in deep meditation when there is no object, no anticipation, no reward on the horizon. It doesn't explain the bliss that arises spontaneously, unbidden, when you're sitting in silence and suddenly the heart opens and love floods through. What if the dopamine is not responding to a person or a memory, but to a frequency — a coherence in the field that the body recognizes and wants to align with?

Oxytocin: The Bonding Hormone

Oxytocin is often called the "cuddle hormone" or the "love hormone." It's released during childbirth, breastfeeding, orgasm, and prolonged physical touch. It promotes bonding, trust, empathy, and social connection.

Studies have shown that oxytocin levels spike when people hug, hold hands, or gaze into each other's eyes. It's also released during acts of kindness, generosity, and care. Oxytocin makes us feel connected as it dissolves the sense of separation between self and other.

But again, oxytocin is a response, not a cause. It's the body's way of saying, "Yes, this connection is real. Lean into it. Trust it. Stay here." Interestingly, oxytocin is also released during meditation, especially loving-kindness (metta) meditation, where practitioners intentionally generate feelings of warmth and compassion toward all beings. No physical touch. No other person. Just the intention to feel love — and the body responds with oxytocin.

This suggests that oxytocin is not just about bonding with others. It's about recognizing coherence, whether that coherence is with another person, with yourself, with the present moment, or with the vast field of Source itself.

Vasopressin: The Commitment Molecule

Vasopressin is less famous than oxytocin, but just as important. It's involved in long-term pair bonding, attachment, and social memory. Prairie voles (small rodents that form monogamous bonds) have high levels of vasopressin receptors in their brains. When scientists block those receptors, the voles lose interest in their partners. When they enhance them, the voles become more devoted.

In humans, vasopressin is associated with the stability of love, the feeling of commitment, loyalty, and enduring connection. It's what turns the intoxicating rush of new love into the steady warmth of long-term partnership. But again, it's a response. Vasopressin doesn't create commitment. It registers it. It's the body's way of locking in a pattern of connection once it's been established.

The Star Player: The D3 Dopamine Receptor

Now let's talk about the receptor that may be the key to love as a sense: the D3 dopamine receptor (Drd3). There are five types of dopamine receptors in the brain (D1, D2, D3, D4, D5), each with a slightly different structure and function. But the D3 receptor is unique. It's not found everywhere. It's highly localized in specific regions of the limbic system including the nucleus accumbens — involved in reward, motivation, and the experience of pleasure, and the islands of Calleja — small clusters of neurons whose function is still mysterious, but which are thought to play a role in processing emotionally salient information.

The D3 receptor is also structurally different from the others. The D2/D3 receptors in the ventral striatum have a higher affinity for dopamine, meaning it's more sensitive; it responds to lower concentrations of dopamine than the other receptors do. And it's involved in sustained states of reward and well-being, not just fleeting pleasure. This is the receptor that lights up during:

Deep meditation
Flow states
Moments of transcendence
The experience of unconditional love

What if the D3 receptor evolved specifically to detect the love frequency? Could it act as the biological interface between the vibrational field and the nervous system, the receptor that says, "Yes, this is it. This is what we've been searching for. This is home"?

If that's the case, then dopamine is not the cause of love; it's the messenger. The D3 receptor detects the coherence in the field, dopamine binds to it, and the brain registers the signal as: "Love is present. Pay attention. Align with this."

Interoception: The Sense of the Internal World

Things get even more interesting with Interoception. Interoception is the sense of the internal state of the body. It's how you know you're hungry, thirsty, tired, or in pain. It's mediated by specialized receptors throughout the body — in the heart, the gut, the lungs, the blood vessels, the skin — that detect things like heart rate, blood pressure, gut motility, breath rhythm, and muscle tension.

These signals travel up the vagus nerve and other pathways to the insula, a deep brain region involved in processing bodily sensations and integrating them with emotions. People with high interoceptive awareness such as those who can accurately sense their own heartbeat, also report more intense emotional experiences. They're more in touch with their feelings, more empathetic, more attuned to subtle shifts in mood and energy.

Here's the key to contemplate. Interoception is not just about sensing your own body. It's about sensing the resonance between your body and the environment. When your heart rate increases, is it because you're anxious? Or is it because you're sensing excitement in the field? When your chest feels warm and open, is it because you ate a good meal? Or is it because you're sensing love?

The body doesn't distinguish. It just sends the signal, "Something is happening. Heart rate up. Breath deeper. Warmth in chest." The brain then interprets that signal based on context, memory, and belief.

What if the experience of love includes a refined form of interoception, not just sensing the state of your own body, but sensing the coherence between your biofield and the larger field? What if, when you "feel love," what you're actually detecting is the resonance — the alignment — the harmonic — between your internal state and something external?

This would explain why love feels embodied. It's not just a thought or an idea. It's a sensation in the chest, the belly, the throat. Love changes your physiology. Heart rate stabilizes. Breathing deepens. Cortisol drops. The parasympathetic nervous system activates.

Love is contagious. When you're in the presence of someone in a coherent state, your body mirrors it. Your heart rhythms synchronize. Your brainwaves entrain. You feel what they're feeling.

Interoception + biofield resonance = the mechanism by which we sense love.

The Heart as Sensory Organ

Let's talk about the heart. For most of history, the heart was considered the seat of emotion, intuition, and love. Then modern science came along and said, "No, the heart is just a pump. All emotion and cognition happens in the brain." But recent research has complicated that view. It turns out the heart is far more than a pump.

The Heart Has Its Own Nervous System. The heart contains approximately 40,000 neurons, a complex network sometimes called the "heart brain." These neurons can sense, process information, make decisions, and even learn independently of the brain.

The heart sends more signals to the brain than the brain sends to the heart (about 80% of the signals along the vagus nerve go up, not down). And these signals influence brain function, affecting perception, decision-making, and emotional processing.

The Heart Generates an Electromagnetic Field. The heart produces the strongest electromagnetic field in the body, about 60 times stronger than the brain's field. This field extends up to 3 feet beyond the body and can be detected by magnetometers.

Remarkably, the pattern of this field changes based on emotional state. During stress, anger, or frustration, the field is chaotic, irregular, jagged. During love, gratitude, or compassion, the field becomes coherent, smooth, sine-wave-like, harmonious.

And here's the kicker: other people can detect your heart's electromagnetic field. Studies have shown that when two people sit close to each other, the heart rhythm of one person can be detected in the brainwaves of the other. This is called cardiac-brain synchronization.

What if this is the mechanism by which we sense love in others? Not through words or facial expressions or body language (though those help), but through the electromagnetic field radiating from their heart? What if the heart is not just a pump, but a receiver and transmitter — a sensory organ that is also tuned to the love frequency?

Heart Coherence and the Love Sense

When the heart is coherent, the entire body shifts. Brainwaves synchronize (especially in the alpha and gamma ranges). The autonomic nervous system balances (sympathetic and parasympathetic in harmony). Immune function improves and cortisol drops. DHEA (the "anti-aging hormone") increases. Decision-making improves and intuition sharpens. This is not just "feeling good." This is a physiological state — a state of optimal function, integration, and alignment. And it's the same state that arises during deep meditation, energy healing, and moments of unconditional love.

Perhaps heart coherence is not the result of love, but the condition that allows love to be sensed more clearly. The heart, when coherent, may simply be tuned in to the love frequency, the way a radio, properly calibrated, picks up a clear signal.

Why the Biochemical Model Is Incomplete

Let's summarize what we know. Dopamine is released when we sense love, but it doesn't explain objectless love. Oxytocin promotes bonding, but it's released even in the absence of another person. Vasopressin supports commitment, but it doesn't explain why we long for love in the first place.

The D3 receptor is localized in reward centers and lights up during transcendent states, but what is it detecting? Interoception allows us to sense internal states, but also resonance with the external field. The heart has its own nervous system, generates a powerful electromagnetic field, and changes its coherence based on emotional state. All of this is real. All of this is measurable. But it's incomplete because it treats love as something the body produces. The context changes if the body doesn't *produce* love.

If the body *detects* love then the biochemistry, the neural pathways, the heart coherence are all just the response to the stimulus. I believe love is not in us, but around us — a field, a frequency, a dimension of reality — and we are simply tuned receivers. This reframe changes everything. If love is a field, we are designed to detect it. We are not lacking love, we are immersed in it, but not always tuned in. We are not creating it through willpower or positive thinking. We are aligning with it, clearing the noise, refining the signal, becoming coherent.

That means the biology is not the end of the story. It's the beginning. It is the interface between the physical and the multidimensional, the body and the field, the trace and the rod.

CHAPTER SEVEN
The Quantum Field and the Nature of Source

This is where we step beyond the laboratory and beyond the microscope. We dive in beyond the neat, measurable world of neurons and neurotransmitters and into the realm where science meets mystery — where the language of physics begins to sound like poetry, and the equations point toward something vast, something conscious, something that feels suspiciously like love. Welcome to the quantum field. Welcome to the nature of Source.

If love is a sense, and if that sense detects a vibrational frequency in the environment, then we need to ask, "What is that frequency? Where does it come from? And what is the field in which it exists?" The answer, I believe, lies at the intersection of quantum physics, consciousness science, and what mystics have always called the unified field — the ground of being, the fabric of reality, the Source from which all things arise and to which all things return.

This is not mysticism dressed up as science. This is the cutting edge of physics itself, pointing toward a reality far stranger, far more interconnected, and far more aware than the mechanistic worldview ever imagined.

The Illusion of Separation

Let's start with the most fundamental insight of quantum physics; there is no such thing as an isolated object. In classical physics, the physics of Newton, the physics we're taught in school, the universe is made of discrete, separate things such as particles, atoms, planets, people. These things interact through forces (gravity, electromagnetism), but they remain fundamentally distinct. You are here. I am there. The table is solid. The air is empty. Everything has clear boundaries. Quantum physics demolished that worldview.

At the quantum level, there are no solid objects. There are only probability waves — clouds of potential, oscillating fields of energy, collapsing into particles only when observed. An electron is not a tiny ball orbiting a nucleus. It's a wave function — a mathematical description of all the places it might be — until something measures it, at which point it "chooses" a location.

It gets even stranger. Particles that have interacted remain entangled. Even if you separate them by light-years, a change to one particle instantly affects the other. It's not through any signal or through any force. Just... because they are still, in some fundamental sense, one thing.

Einstein called this "spooky action at a distance" and hated it. But experiment after experiment has confirmed it. Entanglement is real. Non-locality is real. The universe is not made of separate parts. It is a whole, a seamless, interconnected web of energy and information.

You are not separate from it. You are part of it. Your body is made of atoms that were forged in the hearts of dying stars. Your electrons are entangled with electrons billions of miles away. Your consciousness is not trapped inside your skull — it is interfacing with a field that permeates all of space. Separation is an illusion. And if separation is an illusion, then what we call love, the feeling of connection, of oneness, of dissolving boundaries, is not a delusion. It is recognition of what is true.

The Zero-Point Field: The Fabric of Reality

Let's go deeper. In quantum field theory, the vacuum, the "empty" space between particles, is not empty at all. It's seething with energy. Virtual particles pop in and out of existence. Quantum fluctuations ripple through the void. This is called the zero-point field and is considered the lowest energy state of the quantum vacuum. It is everywhere. Even in the coldest, darkest, most isolated region of space, the zero-point field is there. It cannot be removed. It cannot be turned off. It is the substrate of reality, the background hum from which all matter and energy arise.

What's most remarkable is that the zero-point field is not passive. It's active. It carries information. Some physicists, like Hal Puthoff and

Ervin Laszlo, have proposed that the zero-point field functions as a kind of cosmic memory — an informational matrix in which everything that has ever happened is encoded.

This is not mysticism. This is serious theoretical physics. Laszlo calls it the Akashic field, borrowing a term from ancient Sanskrit that means "ether" or "sky," referring to the subtle substance in which all knowledge is stored. The idea is that every event, every thought, every interaction leaves a ripple in the zero-point field. And because the field is non-local, meaning it connects everything to everything else, all the ripples can be accessed anywhere, anytime, by anything capable of tuning into the field.

This would explain many of the experiences that physics and mainstream science don't want to go near including synchronicity, the "meaningful coincidences" that seem too perfect to be random, and also remote viewing which is the ability to perceive distant locations without sensory input. (Military and adjacent programs around the world are said to use or have used this for spy programs!)

Most of us have experienced strong intuition, the sudden knowing of something you couldn't possibly know through normal channels. This too fits the profile, as does collective consciousness and the hundredth monkey phenomenon, where ideas, inventions, and cultural shifts seem to arise simultaneously in different parts of the world.

It would also explain love. If love is a frequency in the zero-point field, a coherence, a resonance, a tenant, a pattern of information encoded into the fabric of reality itself, then it makes sense that we can sense it, not through our eyes or ears, but through the biofield, through quantum processes in our cells, through the subtle interface between our nervous system and the field.

Quantum Coherence in Biological Systems

For a long time, physicists assumed that quantum effects such as entanglement, superposition, coherence, only happened at extremely cold temperatures, in controlled laboratory conditions. Biology, they said, was too warm, too wet, too noisy for quantum phenomena to survive. They were wrong. In the last two decades, a new field has emerged, entitled quantum biology. And it has discovered that living systems use quantum processes all the time.

Photosynthesis

Plants convert sunlight into chemical energy through photosynthesis. But there's a mystery we didn't learn about in grade school. The efficiency is too high. Nearly 100% of the light energy that hits a chlorophyll molecule gets transferred to the reaction center where it's converted into usable energy. In classical physics, that shouldn't be possible — there should be losses, scattering, wasted energy.

In 2007, researchers discovered that photosynthesis uses quantum coherence, the energy wave explores multiple pathways simultaneously (like Schrödinger's cat being alive and dead at once) and collapses into the most efficient path. This allows plants to be far more efficient than any human-made solar panel.

Bird Navigation

Migratory birds can sense the Earth's magnetic field and use it to navigate thousands of miles with pinpoint accuracy. But the Earth's magnetic field is extremely weak, far too weak to move a compass needle inside a bird's brain. So how do they detect it?

The answer, it turns out, involves cryptochromes — light-sensitive proteins in the bird's retina that form radical pairs when struck by blue light. These radical pairs are quantum-entangled, and the Earth's magnetic field subtly influences the spin states of the electrons, changing the chemical outcome in a way the bird's brain can detect. This is not speculation. This has been experimentally verified. Birds are using quantum entanglement to navigate.

Enzyme Activity

In the realm of biochemistry, it is understood just how much enzymes are the workhorses of biology. They catalyze reactions, break down molecules, build new structures. And they do it fast — far faster than classical chemistry would predict.

Why? Many newer studies show they use quantum tunneling — the phenomenon where a particle "teleports" through a barrier instead of going over it. In enzymatic reactions, protons and electrons tunnel through energy barriers, allowing reactions to happen almost instantaneously. Living systems are quantum systems. Your cells are not just chemical factories. They are quantum processors using coherence, entanglement, and tunneling to function at a level of efficiency and coordination that classical physics cannot explain.

If your cells are quantum processors, then your consciousness, which is described from the coordinated activity of trillions of cells, (which I also personally debate) may also be a quantum phenomenon.

The Orchestrated Objective Reduction (Orch OR) Model

This brings us to one of the most controversial and fascinating theories in neuroscience, the Orch OR model proposed by physicist Roger Penrose and anesthesiologist Stuart Hameroff. The basic idea of the theory is that consciousness does not arise from the connections between neurons (the classical view), but from quantum processes inside neurons, more specifically from inside structures called microtubules.

Microtubules are tiny, hollow tubes that form the structural skeleton of every cell. They're made of repeating protein subunits called tubulins, arranged in a crystalline lattice. And according to Penrose and Hameroff, these tubulins can exist in quantum superposition, meaning multiple states at once, creating a vast network of quantum computation inside every neuron.

When the quantum state reaches a certain threshold, it undergoes objective reduction (OR), a collapse of the wave function, and a moment of consciousness arises. This collapse is not random. It's orchestrated (Orch) by the structure of the microtubules and the larger environment of the brain. And because the collapse is connected to the geometry of spacetime itself (according to Penrose's theory of quantum gravity), consciousness is not in the brain. It's an interface between the brain and the fundamental structure of reality.

Now, the Orch OR model is controversial and many neuroscientists reject it. It has not been disproven, and in fact, recent experiments have found evidence of quantum coherence in microtubules at body temperature which was something that was thought to be impossible.

There is a relevant, compelling part of this theory that aligns to our purpose of better defining love. If consciousness is a quantum phenomenon interfacing with the fabric of spacetime, then it makes perfect sense that we could sense love — a coherence in the field — through quantum processes in our cells. We don't know where the Orch OR model will land with mainstream science but understanding the phenomenon exists, regardless, gives rise to what we experience in our reality. The D3 receptors, the interoceptive pathways, the heart's electromagnetic field may be the classical interface, but beneath them at the quantum level, there may be a deeper resonance, a coherence between the microtubules in your neurons and the zero-point field itself. When that coherence is high, you sense love. When it's low, you feel disconnected.

Fröhlich Coherence: Vibration at the Cellular Level

In the 1960s, physicist Herbert Fröhlich proposed that biological systems generate coherent vibrations. There are oscillations where all the molecules in a system move in sync, like a chorus singing in harmony. These vibrations occur at extremely high frequencies, in the gigahertz range, far above what our ears can hear, but they are real. They have been detected in cell membranes, in proteins, in DNA. And they may play a crucial role in how cells communicate with each other.

What's remarkable is that Fröhlich coherence doesn't require direct contact. It can propagate through the electromagnetic field. Cells can entrain each other and synchronize their vibrations, even when separated by space. This is not woo. This is biophysics. It suggests that living systems are not just chemical machines, but vibrational instruments — oscillators tuned to specific frequencies, capable of resonating with each other and with the larger field.

It is possible that love is one of those frequencies. For this to be true, what would happen is that your cells achieve a certain state of coherence, as they resonate with a frequency in the zero-point field (Source), and one aspect of that resonance is what you experience as love.

The Unified Field: Where Physics Meets Source

Now we arrive at the deepest question. What is the nature of the field itself? Physicists have long sought a unified field theory which is a single framework that unites all the fundamental forces of nature (gravity, electromagnetism, the strong and weak nuclear forces) into one coherent whole. Einstein spent the last 30 years of his life searching for it. He never found it. But many physicists believe it exists, that beneath the apparent diversity of forces and particles, there is a single, unified field from which everything arises.

Some call it the quantum field. Some call it the zero-point field. Some call it spacetime itself. And some, particularly those working at the intersection of physics and consciousness, call it Source.

Source is not a religious concept. It's a scientific hypothesis that says the ground of reality is not inert, not mechanical, not dead, but alive, intelligent, aware. Most importantly is says that consciousness is not a byproduct of matter, but a fundamental property of the field itself.

This is the view of panpsychism — the idea that consciousness is woven into the fabric of reality at every scale, from electrons to galaxies. It's the view of idealism — the philosophical position that mind, not matter, is primary. It's the view of the mystics, who have always said, Consciousness is not in you. You are in Consciousness.

If consciousness is the ground of being, if the unified field is aware, then what we call love may simply be the natural state of that field. It would not be something added on, not something earned. *Love would be a default frequency, a coherence, the harmony, the resonance that exists when nothing is distorting it.*

When your nervous system is calm, your heart is coherent, your cells are oscillating in sync with the zero-point field, what you sense is not a new thing. You sense what was always there. You sense Source. And Source feels like love, due to the receptors in our bodies.

The Dimensional Trace, Revisited

Let's return to the metaphor that began this book: the rotating rod and the U-shaped trace. The rod represents a higher-dimensional reality — something vast, coherent, whole. The plexiglass plane represents our limited perception, the 3D world we navigate with our five senses. And the U-shaped trace represents what we can detect when the rod intersects our plane.

Love is that trace, likely one trace of many. It's the shadow cast by something immense, something multidimensional, something that exists at a deeper level of reality than our biology can fully grasp. But we can sense it through the biofield, through quantum coherence, through the resonance between our cells and the zero-point field. And when we do, when the angle is right, when the noise quiets, when the system aligns, we don't just feel good. *We remember what we are.*

We remember that we are not isolated, separate, alone. We remember that we are waves in an ocean, notes in a symphony, threads in a tapestry. We remember that the boundary between self and other, between inside and outside, is not as solid as it seems. We remember that we are held by the field, by Source, by the unified coherence that holds all things together. That remembering is love.

The Science of the Heart, Revisited

Now, a word about HeartMath. They've shown that the heart's electromagnetic field is measurable, coherent, and responsive to emotional state. Their data points to something real — the heart is not just a pump. It's a transmitter. When it's coherent, it broadcasts a signal, a frequency, that others can detect.

We may further consider, from all the information gathered so far, that the heart is not just transmitting its own signal. It's also potentially resonating with the zero-point field. It's tuning into a frequency that was already there. And when it does, the brain registers that resonance as *love is present.* This is not HeartMath's theory. This is a quantum, dimensional, Source-based model. But the heart coherence research supports it.

If love is a frequency in the unified field — if it's a coherence encoded into the zero-point field, accessible through quantum processes, detectable through the heart and the biofield, then:

Love is not scarce. It's infinite. It's the ground state of reality. Love is not conditional. It doesn't depend on worthiness, effort, or behavior. It simply is. Love is not personal. It's transpersonal. It flows through us, not from us. Love is not an accident. It's the architecture of being. It's what holds the universe together.

The reason we long for love is not because we're broken. It's because we're tuned instruments that are designed, evolved, optimized to detect it. We are the plexiglass plane, sensing the rod. We are the radio, tuned to the broadcast. We are the wave, recognizing the ocean. When we sense it clearly — when the signal comes through undistorted — we don't just feel love. We are love.

CHAPTER EIGHT
Plants, Animals, and the Love Frequency

If my theory is correct and love is a sense, a faculty by which the body detects an external stimulus, then we need to ask the next question. Is it uniquely human?

We tend to think of love as requiring sophisticated biology including brains capable of processing emotion, nervous systems that produce oxytocin, limbic regions that light up on fMRI scans. We imagine love perception as the pinnacle of evolutionary development, something only the most complex organisms can experience. Do you need a brain to sense love? Do you need neurons, synapses, a limbic system? Do you need dopamine receptors and oxytocin pathways and a prefrontal cortex capable of abstract thought? Or is love something deeper, something more fundamental, a vibrational frequency that all living systems can detect, regardless of their biological complexity?

The answer, I believe, lies not in the laboratory, not in the fMRI scanner, but in the garden. In the forest. In the quiet intelligence of beings we have long dismissed as passive, unconscious, mechanical.

It lies in plants. Because if plants can sense love — and the evidence suggests they can — then love cannot be neurological. It cannot be psychological. It cannot be a trick of evolution designed to keep humans bonding and reproducing. It must be something else entirely. That something would be vibrational, universal and woven into the fabric of life itself. If love exists as a fundamental frequency of reality it would be something so basic that even the simplest forms of matter and life respond to it (though I hate to call plants simple).

The Astonishing Awareness of Plants

For most of human history, we have treated plants as background. Decoration. Resources. We have cut them, eaten them, burned them, transplanted them, and never once considered that they might be aware. But they are.

Plants don't have brains and they don't have neurons. They don't have a central nervous system. Yet they perceive their environment with astonishing sophistication. They sense light, gravity, temperature, moisture, chemicals, vibrations, and even the presence of other organisms. They communicate with each other through underground fungal networks, vast, interconnected webs jokingly called the "wood wide web", sharing nutrients, sending distress signals, and cooperating in ways that look suspiciously like mutual aid.

They remember. A plant that has been dropped will close its leaves in anticipation the next time it's picked up. A plant that has been attacked by a specific insect will produce defensive chemicals the next time it detects that insect's scent, even years later.

They make decisions. Roots grow toward water. Vines climb toward light. Some plants can "choose" between two sources of nutrients based on which is more abundant or more easily accessible. And here's the most of us never thought about — they respond to intention, not just to the physical stimuli we can measure that includes light, water, touch, but to something subtler. It's a little something we might call awareness. Or presence. Or love.

Cleve Backster and Primary Perception

In 1966, a lie detector expert named Cleve Backster made a discovery that would change the way some people think about plants forever. Backster was working late in his New York office, watering his dracaena plant, when a thought occurred to him: What would happen if I attached a polygraph to a plant leaf? Polygraphs (lie detectors) measure galvanic skin response, the subtle electrical changes in the skin caused by emotional arousal. When a person lies, their stress response activates, and the polygraph detects it. But a plant has no skin. No nervous system. No emotions, according to conventional biology.

Still, Backster was curious. He attached the electrodes to a leaf and watched the readout. Nothing unusual. Just a slow, steady tracing. Then, on impulse, he thought, "What if I burned the leaf?" He didn't do it. He only thought it. But the moment the thought formed in his mind, the polygraph needle spiked with a sharp, dramatic reaction, as if the plant had detected his intent.

Backster was stunned. He tried it again. Same result. The plant seemed to be responding not to a physical threat, but to his intention, a thought, his mental state, his focus, his awareness. He called this phenomenon primary perception — the idea that all living cells possess a form of awareness, a sensitivity to the thoughts and emotions of other beings, even across distance. Over the next several years, Backster conducted hundreds of experiments. He found that plants responded to:

The death of nearby organisms. When he dropped live brine shrimp into boiling water in another room, the plant registered a response, a spike in electrical activity at the exact moment the shrimp died.

The emotional state of their caretakers. Plants seemed to "know" when their owners were happy, sad, stressed, or calm, even when the owner was miles away.

The intention to harm them. Not the act itself, but the thought, the decision, a shift in awareness.

Backster's work was controversial. It still is. Mainstream science dismisses it as pseudoscience, flawed methodology, wishful thinking. And to be fair, the experiments were difficult to replicate under controlled conditions. The phenomenon was subtle, inconsistent, dependent on variables that were hard to isolate.

However, here's what's undeniable. Something was happening, and it wasn't explainable by conventional biology. If plants were responding to intention, to awareness, to the emotional field of another being, then they were detecting something that has no physical form, something that doesn't travel through sound waves or light waves or chemical molecules. They were detecting a vibrational field, a thought, coherence, a frequency.

Sound, Vibration, and the Language of Plants

In recent years, the scientific study of plant intelligence has exploded. And while most researchers avoid the word "love," they are discovering that plants are exquisitely sensitive to vibration.

Plants respond to sound. Studies have shown that plants grow faster and healthier when exposed to certain types of music, particularly classical music and frequencies in the range of 125-250 Hz. Root growth accelerates. Seed germination improves. Chlorophyll production increases. Conversely, plants exposed to harsh, chaotic sounds including heavy metal, white noise, aggressive frequencies show stunted growth, irregular patterns, and signs of stress.

Why? Because sound is vibration and vibration is information. Plants don't have ears, but they have membranes, cell walls, proteins that oscillate in response to mechanical waves. They are vibrational instruments, tuned to the frequencies around them.

Plants respond to human voice. In a famous series of experiments, researcher Dorothy Retallack found that plants thrived when spoken to with kindness, gentleness, and intention. They leaned toward the speaker. Their leaves expanded. Their growth accelerated. But when spoken to with anger, criticism, or neglect, the plants leaned away. Their growth slowed. Some even died.

This wasn't about the words. Plants don't understand English. It was about the frequency — the vibrational quality of the voice, the emotional state of the speaker, the coherence of the field being transmitted.

Plants respond to intention. In the 1970s, researcher Marcel Vogel (a senior scientist at IBM) conducted experiments showing that plants responded to focused intention. When he held a clear, loving intention toward a plant, silently, without touch, without sound, the plant's electrical activity changed. It became more coherent, more rhythmic, more alive. When he held a neutral or negative intention, the plant's activity became chaotic, erratic, stressed.

Vogel's conclusion was radical. He said plants are not responding to physical stimuli alone. They are responding to consciousness, to awareness, to the quality of attention being directed toward them. And if that's true, then plants are sensing something we can't see, hear, or measure with conventional instruments. They are sensing the field and its coherence which includes the love frequency.

Biophotons: The Light Within

Here's where it gets even stranger. All living cells emit light, not visible light but ultra-weak photon emissions in the UV and infrared range, far too faint for the naked eye to detect. But they're real. They've been measured. And they're called biophotons. Biophotons are thought to be a byproduct of metabolic processes, the natural "glow" of biochemical reactions happening inside cells. Some researchers believe they may serve a deeper function of communication.

German biophysicist Fritz-Albert Popp spent decades studying biophotons and proposed that they act as a signaling system within and between organisms. Cells use light to communicate with each other, coordinating activity, synchronizing rhythms, sharing information instantaneously across the body.

The coherence of biophoton emissions changes based on the state of the organism. Healthy, thriving cells emit coherent light — organized, rhythmic, laser-like. Diseased or stressed cells emit chaotic light — scattered, irregular, disorganized. Plants emit biophotons. Animals emit biophotons. Humans emit biophotons. And when two living systems come into proximity, their biophoton emissions can entrain and synchronize, resonate, align. This is not metaphor. This is biophysics.

What is light? A waveform. What other types of waveforms make up Source? Unknown. Yet what if love is biophotonic type coherence? What if, when we say we "feel love" in the presence of another being, what we're actually detecting is the coherence of their 'light' field aligning with ours? What if plants, with no brain and no nervous system, are sensing that same coherence and responding to it?

Fröhlich Coherence: The Hum of Life

In the 1960s, physicist Herbert Fröhlich proposed that living systems generate coherent vibrations, oscillations at specific frequencies, where all the molecules in the system move in sync, like a choir singing in perfect harmony.

These vibrations occur at extremely high frequencies in the gigahertz range (billions of cycles per second), far above what we can hear. But they are real. They have been detected in cell membranes, in proteins, in DNA.

Fröhlich suggested that these coherent oscillations play a crucial role in how cells communicate with each other. They don't need direct contact. They don't need chemical messengers. They can resonate by tuning into each other's frequency and coordinating their behavior at a distance. This is called Fröhlich coherence, and it's been observed in living tissues across the biological spectrum including bacteria, plants, animals, humans. What's remarkable is that coherence doesn't require consciousness (as we conventionally define it). A single-celled organism can oscillate coherently. A plant can oscillate coherently. A protein molecule can oscillate coherently.

Here's the key: coherence can be influenced, disrupted, enhanced. When a system is stressed such as exposed to toxins, deprived of nutrients, attacked by pathogens, its coherence breaks down. The oscillations become chaotic, irregular, dissonant. When a system is thriving, nourished, safe, in resonance with its environment, its coherence increases. The oscillations become smooth, rhythmic, harmonious.

Love may enter the picture here. Assume love is the frequency that promotes coherence. What if, when we direct love toward a plant, an animal, another person, we are not projecting an emotion, but transmitting a coherent field, a vibrational pattern that the other organism's cells can detect and entrain with?

What if love is not what we feel, but what we broadcast — and what others sense?

Animals and the Love Field

If plants can sense love, what about animals? Anyone who has lived with a dog, a cat, a horse, a bird knows the answer. Animals sense emotional states with uncanny accuracy. They know when you're sad. They know when you're anxious. They know when you're lying. For a long time, we assumed this was body language and subtle cues in posture, facial expression, scent. That's definitely part of it. And there's something more.

Dogs have been shown to synchronize their heart rhythms with their owners. When a person and their dog sit quietly together, their heart rate variability (HRV) patterns align, not through touch, not through sound, but through the electromagnetic field radiating from the heart.

Horses respond to human heart coherence. Studies in equine-assisted therapy have found that when a person shifts into a coherent heart rhythm through breathing, meditation, or emotional regulation, the horse's nervous system responds within seconds. The horse relaxes and becomes more present. It mirrors the coherence.

Dolphins and whales communicate through sound, and also through bioelectric fields. Their brains are extraordinarily large, and their capacity for empathy, play, and social bonding rivals (and may exceed) that of humans. They have been known to approach humans in distress and offer what can only be described as comfort, meaning a presence, a resonance, a field of calm.

What are they sensing? Not words. Not even body language, in many cases. They are sensing the field. The coherence. The frequency we are broadcasting. And when that frequency is love — when the heart is coherent, the nervous system calm, the mind quiet — they respond. They resonate. They align.

Love Is Not Neurological

Please contemplate this. If plants and animals, beings without a human brain, without a prefrontal cortex, without the neural architecture we associate with emotion, can sense and respond to love, then love cannot be a product of the brain.

It cannot be dopamine, oxytocin, vasopressin, or any other neurochemical. Those chemicals may be involved in how humans experience love, but they are not the source of it.

Love must be something that exists outside the body — a field, a frequency, a vibrational pattern that all living systems can detect and respond to, each in their own way.

Plants respond with growth, health, coherence. Animals respond with trust, bonding, attunement. Humans respond with dopamine surges, heart coherence, brainwave synchrony — but also with something deeper which includes the felt sense of connection, expansion, homecoming. The stimulus is the same, the mechanism of detection varies, but the response is universal.

Life recognizes love. Life resonates with love. Life moves toward love.

The Implications Are Staggering

If love is a universal vibrational frequency detectable by all living systems, not just humans, then:

Love is not a human invention. It's not cultural, not psychological, not a trick of evolution. It's fundamental. It's woven into the structure of life itself.

Love is not scarce. It's not something you have to earn, find, or deserve. It's a field you can tune into, and the more coherent you become, the more clearly you sense it.

Love is not subjective. Just as light is light whether or not you see it, love is love whether or not you recognize it. It's measurable. It's consistent. It's real.

Love is not sentimental. It's not soft, passive, or weak. It's a force, a coherence that organizes systems, promotes health, facilitates communication, and drives living organisms toward connection and growth.

Love is the signal, not the noise. The longing for love, the ache of separation, the relentless human search for connection are not signs of dysfunction. They are signs of a system designed to detect and align with a specific frequency. We are tuning forks, resonating with a field that has always been there.

Following the Trace

Remember the rotating rod and the U-shaped trace? The two-dimensional plane could only perceive a slice of the three-dimensional rod. It couldn't see the whole. But the trace was a clue, a hint that something larger was intersecting its reality. Love is that trace.

The fact that plants sense it, that animals sense it, that all living systems respond to it in measurable, repeatable ways tells us something profound: The rod is real. The higher-dimensional reality is real. And we are all living in it, sensing it, responding to it, even when we don't fully understand what it is.

So what is love? It's not an emotion. It's not a neurochemical reaction. It's not a psychological construct. It's a frequency. A coherence. A vibrational field that all life can detect. And the reason we long for it is not because we lack it. It's because we sense it the way a plant senses sunlight, the way a bird senses magnetic north, the way a fish senses the current.

We are wired for love. Tuned to it. Drawn toward it. Because it's not just what we feel. It's what we are.

CHAPTER NINE
Love Beyond Emotion

We have a language problem. We use the same word — love — to describe wildly different experiences:

- The rush of attraction when you meet someone new
- The ache of missing someone who's gone
- The warmth you feel holding a sleeping child
- The gratitude that wells up watching a sunset
- The peace that floods through you in deep meditation
- The fierce protectiveness you feel for someone in danger
- The quiet contentment of a long partnership
- The expansive bliss that arises when you touch Source

These are not the same experience and yet, we call them all "love." No wonder we're confused. No wonder science can't pin it down. No wonder philosophers argue about definitions. No wonder people say things like, "I don't even know what love is anymore."

The problem is not that love is mysterious or indefinable. The problem is that we've been conflating three fundamentally different phenomena of emotions, feelings, and sensing. Until we untangle them, we can't see love clearly.

This chapter is that untangling. It's a map of the terrain, showing where emotions end, where feelings begin, and where sensing arises. It's here in the realm of sensing, that we finally encounter love in its purest form, not as something we do, not as something we feel about something, but as something we perceive — a frequency, a field, a dimensional signal intersecting our awareness.

Emotion: The Body's Reaction

Let's start with emotion. In neuroscience, an emotion is a rapid, automatic, physiological response to a stimulus. It's what happens in your body before you consciously register what you're experiencing. You see a snake. Your amygdala fires. Adrenaline floods your system. Your heart rate spikes. Your muscles tense. You freeze or run or strike. That's fear — the emotion. It happens in milliseconds, below the threshold of conscious awareness. You don't decide to be afraid. Your body reacts, and only afterward do you realize what happened.

The same is true for all the basic emotions including fear, anger, disgust, surprise, sadness, joy. These are hardwired survival responses, evolutionarily ancient, designed to help you navigate threats and opportunities in the environment. Emotions are fast, automatic, and embodied. They involve the amygdala (threat detection, emotional salience), the hypothalamus (triggering hormonal responses), the autonomic nervous system (ramping up or calming down the body), and the brainstem (regulating basic survival functions).

Emotions are not "in your head." They are in your body and are a cascade of neurochemical and physiological changes preparing you to act. Emotions are reactions to stimuli. They are about something such as a perceived threat, a perceived opportunity, a perceived loss.

When we say "I feel love" in the sense of an emotion, what we're usually describing is excitement (when you see the object of your affections), attachment anxiety (when you fear losing them), relief (when they return after an absence), desire (when you're attracted to someone), and protectiveness (when someone you care about is threatened).

These are real. These are valid. But they are reactions triggered by the presence, absence, memory, or anticipation of a person or situation. This is not love itself. This is your body responding to love, or to the trace of love, the promise of love, the memory of love.

Feeling: The Mind's Interpretation

Let's talk about feelings. A feeling is what happens when emotion becomes conscious. It's the story you tell yourself about what the emotion means.

You feel your heart racing. Your body is tense. You're breathing quickly. Is that fear? Excitement? Anxiety? Anticipation? Your brain doesn't know yet. It has to interpret the bodily signals based on context, memory, belief, and expectation. And that interpretation is what we call a feeling.

Neuroscientist Antonio Damasio makes this distinction beautifully. *Emotions happen in the body. Feelings happen in the mind.* Emotions are automatic. Feelings are constructed. This is why the same physiological state can be interpreted in completely different ways such as a racing heart at the top of a roller coaster = exhilaration, a racing heart before a speech = anxiety, and a racing heart when you see your crush = infatuation. These are the same body state but with different feeling because feelings are shaped by context (Where am I? What's happening?), memory (What happened the last time I felt this way?), belief (What does this sensation mean?), and language (What word do I use to describe this?).

Feelings involve the insula (integrating bodily signals with awareness), the anterior cingulate cortex (monitoring emotional states), and the prefrontal cortex (creating narratives, making meaning). Feelings are slower than emotions. They require reflection, interpretation, storytelling. And because they're constructed, they can be inaccurate.

You might feel unloved, even when you are surrounded by people who care for you because your interpretation of their behavior is filtered through old wounds, insecurities, or beliefs about your worthiness. You might feel love for someone who is harmful to you, because your nervous system has learned to associate certain patterns (intensity, unpredictability, abandonment) with connection. Feelings are interpretations. And interpretations can be wrong.

Sensing: Direct Perception of the Stimulus

Now we come to the third layer of sensing. Sensing is what happens before emotion and feeling. It's the direct perception of a stimulus in the environment, meaning the raw data your body is detecting before it gets filtered, interpreted, or stormed over by the mind.

When you see a red apple, your photoreceptors are detecting light waves at approximately 650 nanometers. That's sensing. The fact that you perceive it as "red" and interpret it as "apple" happens later, in the brain's visual processing centers. But the initial detection, the moment the light hits the retina and the photoreceptor responds, that's pure sensing.

When you hear a sound, your cochlear hair cells are detecting vibrations at a specific frequency. That's sensing. The fact that you recognize it as a voice, a song, an alarm is interpretation. But the initial transduction, the moment the hair cell bends and sends an electrical signal, is sensing. Sensing is stimulus detection. It's the interface between the external world and your internal awareness.

Sensing happens without emotion or feeling. Those come later. You can sense light without feeling anything about it. You can sense sound without having an emotional reaction. You can sense pressure on your skin without attaching meaning to it. Sensing is neutral. It's pure perception. It's the body saying, "Something is here. I detect it." This is where love appears to first live, not in the emotion, not in the feeling, but in the sensing. We add those later.

Love as Stimulus, Not Response

Let's apply this framework to love.

Emotion: Your heart races when you see your partner. You feel excited, anxious, attached. This is your body reacting to the stimulus of their presence (or absence, or memory).

Feeling: You think, "I love them. I need them. I'm afraid of losing them. I hope they love me back." This is your mind interpreting the bodily signals, weaving them into a story shaped by past experience and belief.

Sensing: Beneath all of that, and in ways we are not trained to tune into, beneath the racing heart, beneath the story, there is a direct perception of something present in the field. It's a coherence and a resonance and a vibrational frequency that your biofield is detecting and your body is responding to. That is love. It's not the emotion or the feeling. It's the stimulus itself, the thing that triggered the cascade in the first place. You can sense love without any emotional or cognitive overlay. This is what happens in deep meditation.

Meditation: Bypassing the Loop

When you sit in meditation, once the mind is truly still, truly quiet, so that the 'constructed self' recedes, something remarkable happens. The emotional loop quiets down. The amygdala stops scanning for threats. The hypothalamus stops triggering stress hormones. The autonomic nervous system shifts from sympathetic (fight-or-flight) to parasympathetic (rest-and-digest). The feeling loop quiets down. The prefrontal cortex stops generating narratives. The inner voice softens. The mind stops interpreting every sensation as something that needs analysis, fixing, or response.

In that stillness, what remains? Pure sensing. The body is attuned to the field. The heart is oscillating in coherence. The cells are resonating at their natural frequency. The biofield is open and receptive. You are no longer emitting, but receiving. And in that openness, love spontaneously arises, not as an emotion about someone, not as a feeling toward something, but as a presence. It is a warmth, an expansion, a boundless, objectless sense of connection, safety, and homecoming. This is not romantic love. This is not attachment. This is not desire. This is not the neurochemical cascade triggered by seeing a face, receiving a text, or anticipating a reunion. This is unconditional love — love without an object, love without a reason, love that simply is. It arises because, in the absence of mental noise and emotional reactivity, you are finally perceiving the stimulus directly. You are sensing the field.

Why Unconditional Love Is Not an Emotion

Here's where the conventional model breaks down. Science tells us that love is dopamine, oxytocin, vasopressin. It's attachment. It's bonding. It's an evolutionary strategy to keep us mating, parenting, cooperating. But that model cannot explain unconditional love. In deep meditation, there is no person to bond with, no face to trigger dopamine, no body to touch and release oxytocin, no evolutionary advantage to sitting still with your eyes closed, feeling boundless love for all beings.

And yet, it happens. Consistently. Across cultures. Across traditions. Across individuals who have never met, never trained together, never even heard of each other. Buddhist monks in Tibet experience it. Sufi mystics in Persia experience it. Christian contemplatives in medieval Europe experienced it. Modern meditators in New York, Mumbai, São Paulo experience it. And I experienced it too, as the ignition point which led me down this *love* rabbit hole in the first place, because it was prevalent, predominant, and curious enough for me to contemplate and create a new theory about it.

The descriptions are remarkably consistent. First comes a sense of expansion, as if the boundaries of the self dissolve. Next, a warmth in the chest, often described as "the heart opening" and the state of profound peace, as if all seeking has stopped. Soon comes a feeling of being held, loved, safe even in the absence of any external person or presence appears, and the recognition of this is what I've been searching for. This is home.

This cannot be an emotion. Emotions require an object and something to react to. This is objectless. This cannot be a feeling. Feelings require interpretation. This is pre-cognitive as it arises before thought. This is sensing, the direct perception of a stimulus in the field, the body detecting a frequency and signaling the neurochemical pathways to create a response. It's the biofield resonating with coherence, and the heart tuning into the love that was always there.

The Three Layers, Visualized

Let's map it:

Sensing = Direct perception of stimulus → Biofield, receptors, cells detecting coherence in the field → heart resonance; biophotonic alignment

Emotion = Automatic bodily reaction → Amygdala, hypothalamus, autonomic nervous system → Heart racing, warmth in chest, tears, oxytocin release

Feeling = Conscious interpretation → Insula, anterior cingulate, prefrontal cortex → "I love them," "I feel connected," "I'm afraid of losing this"

Most of the time, we experience all three layers simultaneously. You sense love in the field, your body reacts emotionally, and your mind creates a feeling/story about it. But they are not the same. And in meditation or in the presence of a highly coherent being, or during energy healing, or in moments of grace, the emotional and cognitive layers can drop away, leaving only the sensing. That's when you experience love in its purest form.

Why This Distinction Matters

Understanding the difference between sensing, emotion, and feeling changes everything. It explains why love can feel so different in different contexts. Romantic love (emotion + feeling) is intense, personal, and often turbulent because it's filtered through attachment, desire, fear of loss.

Unconditional love (pure sensing) is calm, spacious, stable because it's not about anyone or anything. It's the field itself. It explains why we can love people who hurt us. The sensing of love (the coherence in the field) can be real, even when the feeling (the interpretation) is distorted by trauma, conditioning, or unmet needs. You can sense love and still be in an unhealthy relationship because your nervous system has learned to interpret intensity, chaos, or abandonment as "connection."

It explains why meditation and spiritual practice are so transformative. They don't create love. They clear the noise so you can sense the love that was always there. They quiet the emotional reactivity and the mental storytelling, allowing the biofield to attune to the frequency of the field itself. This reframes the search. We are not searching for a person who will make us feel loved. We are not searching for an experience that will trigger the dopamine rush. We are searching for alignment with the stimulus itself — the coherence, the resonance, the frequency of love in the field. The reason we keep searching in relationships, in achievements, in substances, in experiences is because we keep mistaking the emotion and the feeling for the sensing.

We think, "If I can just find the right person, the right situation, the right combination of circumstances, I'll feel this love all the time." But that's backwards. The love is not in the person. The love is in the field. And the person, when they are coherent, when their biofield is aligned, simply allows you to sense it more clearly.

Love Is the Stimulus

Here's the shift and a new paradigm. The old model says love is an emotion we generate in response to a person or situation. Our new model reveals love is a stimulus in the field that we sense and our body responds with emotions, and our mind responds with feelings, but the love itself is not the response. It's what we're responding to.

Just as light exists whether or not you open your eyes, love exists whether or not you sense it. Just as sound exists whether or not you listen, love exists whether or not you tune in. Love is not produced by the brain. Love is detected by the body. And the more coherent you become — the quieter your nervous system, the more aligned your biofield, the more open your heart — the more clearly you sense it. This is not mysticism. This is sensory science applied to a stimulus we've been misunderstanding for millennia.

The Implications

If love is a stimulus — a frequency in the field — then you are not wrong for wanting it so badly. You are a sensory instrument, designed to detect it. The longing is not a deficiency; it's a homing signal. You don't have to earn love. You can't earn the ability to see light or hear sound. You just open your eyes. You just listen. Love works the same way. You align. You attune. You open. It's always available. The field is constant. Your access to it fluctuates based on the state of your nervous system, the coherence of your biofield, and the clarity of your awareness but the love itself never leaves.

Other people can help you sense it. When you're in the presence of someone who is coherent to your own frequency there is only harmony and your own biofield can entrain. You resonate. Everything relaxes. You tune in. This is not dependency. This is resonance. It's even stronger when someone whose biofield is aligned with the love frequency, and you entrain.

You can cultivate love. Through meditation, breathwork, energy work, and coherence practices, you can train your system to detect the love frequency more clearly, not to create it but to sense it.

The Journey Home

Remember the U-shaped trace? The rotating rod passing through the plexiglass plane?

For most of your life, you've been looking at the trace — the emotions, the feelings, the fleeting experiences of connection and loss — and thinking, this is love. This is what I'm searching for. But the trace is not the rod. The shadow is not the object. The ripple is not the wave.

Love is the rod. Love is the object. Love is the wave. It's vast, coherent, and multidimensional. It's always rotating, always intersecting your reality, always leaving traces you can sense, if you know how to look. And now you do.

CHAPTER TEN
The Vibrational Nature of Love

Everything vibrates. This is not metaphor. This is not New Age wishful thinking. This is physics. At the quantum level, there are no solid objects. There is only energy — oscillating, pulsing, vibrating at different frequencies. The chair you're sitting on is not solid. It's a dense cloud of atoms, and those atoms are mostly empty space, held together by electromagnetic forces that are themselves vibrations in the quantum field.

Your body is not solid. It's a symphony of oscillations, of electrons orbiting nuclei, molecules spinning and tumbling, proteins folding and unfolding, cells pulsing with electrical and chemical activity, organs resonating at specific frequencies. Even your thoughts are vibrations, patterns of electrical activity propagating through neural networks, creating waves that can be measured with an EEG.

Everything that exists — matter, energy, consciousness — is vibration. And if love is real, if it's not just a concept, not just a chemical reaction, not just a psychological construct, then it, too, is likely to be vibration.

This chapter is about understanding love as frequency, as a specific vibrational pattern in the quantum field that all living systems can detect and entrain with. This is where science becomes poetry and where equations become experience. It's where the abstract becomes tangible. Welcome to the vibrational nature of love.

What Is Frequency?

A frequency is the number of oscillations per unit of time. It's how fast something vibrates. Sound waves vibrate at frequencies we can hear (20 Hz to 20,000 Hz). Light waves vibrate at frequencies we can see (430 trillion Hz to 750 trillion Hz). Radio waves, X-rays, gamma rays are all vibrations at different frequencies. Your heart beats at a frequency of about 1 Hz (once per second). Your brainwaves oscillate at frequencies between 0.5 Hz (deep sleep) and 100 Hz (high-level cognitive processing). Your cells vibrate at gigahertz frequencies (billions of cycles per second).

Different frequencies carry different information. Change the frequency, and you change what's being transmitted. This is how a radio works. The air around you is filled with radio waves with dozens, maybe hundreds of stations broadcasting simultaneously. You only hear one at a time, because your radio is tuned to a specific frequency. Change the dial, and you access a different broadcast. The stations were always there. You just weren't tuned in.

Does consciousness work the same way? Is the field around you filled with frequencies — emotions, thoughts, intentions, states of being — and your nervous system acts as a tuner, picking up the frequencies you're aligned with and rejecting the ones you're not aligned with? What if love is one of those frequencies? Not the only frequency, but a specific vibrational pattern a specific coherence, a harmony, a resonance that, when detected, creates the experience we call love?

What Is Coherence?

A frequency alone is just oscillation. When oscillations align, meaning when they move together in sync, reinforcing each other, that's coherence. Think of a choir. When everyone is singing their own note at random, it's noise. But when everyone sings in harmony with same rhythm, complementary pitches, synchronized breath, it becomes music. That's coherence.

Or think of light. A flashlight emits incoherent light, photons scattered in all directions, at slightly different wavelengths, out of phase with each other. But a laser emits coherent light where all the photons are moving in lockstep, same wavelength, same phase, amplifying each other. That's why a laser can cut through steel, while a flashlight just illuminates a wall.

Coherence is alignment. Coherence is power. Coherence is order emerging from chaos. And in biological systems, coherence is health. When your heart is coherent, the rhythm is smooth, regular, sine-wave-like. When it's incoherent, the rhythm is jagged, irregular, chaotic. Studies have shown that heart coherence is associated with lower stress hormones (cortisol, adrenaline), improved immune function, better decision-making and cognitive performance, enhanced emotional regulation, and increased feelings of well-being, gratitude, and — yes — love.

When your brainwaves are coherent, oscillating in sync across different regions of the brain, you experience deep meditation states, flow States, peak performance, mystical experiences and the dissolution of self-other boundaries. Coherence is not just "feeling good." Coherence is the state in which the system is optimally organized, efficiently communicating with itself, and resonating with the larger field.

Here's the key. Coherence can be transmitted. When you're in a coherent state, your electromagnetic field becomes coherent. That coherent field can influence other systems around you, entraining them, synchronizing them, bringing them into resonance. This is why being in the presence of a calm, centered person makes you feel calm. This is why meditating in a group feels different than meditating alone. This is why a baby stops crying when held by someone whose nervous system is regulated. Coherence spreads. Coherence resonates. *Coherence is contagious.*

Love, I propose, is the experience of coherence, both internal (your own system coming into alignment) and external (resonating with a coherent field outside yourself).

What Is Entrainment?

Entrainment is the phenomenon where two oscillating systems synchronize with each other. The classic example is metronomes on a moving platform. If you place several metronomes on a table, each ticking at its own pace, they'll remain out of sync. But if you place them on a platform that can move slightly, a board resting on soda cans for example, something remarkable happens. Within a few minutes, all the metronomes synchronize. They start ticking together, in perfect rhythm, as if they're communicating.

They're not synching through sound or touch. They're communicating through vibration. The platform transmits the oscillations, and the metronomes entrain, adjusting their frequencies to perfectly match each other. This happens everywhere in nature. Fireflies flash in unison across entire forests. Crickets chirp in synchrony. Women living together often synchronize their menstrual cycles. Heart cells in a petri dish, initially beating at different rates, will synchronize and beat as one.

Entrainment is not conscious and it's not intentional. It's automatic, and a fundamental property of oscillating systems. It also happens with biofields. When two people sit close together, their heart rhythms can entrain. Studies have shown that a mother's heartbeat can synchronize with her baby's, a therapist's heart rhythm can influence a client's, and two people in deep conversation or meditation can synchronize brainwaves and heart rhythms without even realizing it. This is not metaphor. This is measurable. You can detect it with an (ECG) electrocardiogram or an electroencephalogram (EEG).

When you're in the presence of someone whose biofield is coherent, your biofield can entrain to theirs. You synchronize. You resonate. You align. Importantly, if their biofield is coherent at the frequency of love, if they are deeply at peace, deeply open, deeply aligned with Source, then being near them allows you to sense that frequency more clearly.

This is what happens in the presence of enlightened beings. This is what happens in energy healing sessions. This is what happens when you sit in meditation with a teacher whose field is so stable, so coherent, that just being near them quiets your nervous system and opens your heart. They are not giving you love. They are resonating at the frequency of love and your system is entraining to it.

The "Love Frequency": 528 Hz

Some researchers have proposed that love has a specific frequency of 528 Hz. This is one of the so-called "Solfeggio frequencies", a set of tones used in ancient Gregorian chants, said to have healing properties. The 528 Hz tone is often called the "Love Frequency" or the "Miracle Tone," associated with DNA repair, transformation, and heart coherence.

Is this scientifically validated? Not at all. The research is sparse, the studies are small, and the claims are often inflated by people selling sound healing products. What's interesting is that 528 Hz is close to the natural resonance of many biological systems. It's in the range where sound can influence cellular activity, brainwave patterns, and heart rhythms.

Some studies have shown that exposure to 528 Hz reduces stress and increases relaxation. Plants exposed to 528 Hz show increased chlorophyll production and growth. Water exposed to 528 Hz (in the famous Masaru Emoto experiments, which we'll discuss shortly) forms more symmetrical, aesthetically pleasing ice crystals.

Does this mean love literally vibrates at 528 Hz? No, not likely. It's not likely love is a single frequency and it is more likely love is a pattern of frequencies, a harmonic, a coherence across multiple scales. But the idea that love has a vibrational signature, a resonance that can be detected, transmitted, and entrained with is not far-fetched. It's consistent with everything we know about biofields, coherence, and entrainment.

Masaru Emoto and the Memory of Water

No discussion of vibration and consciousness would be complete without mentioning Masaru Emoto and his controversial water crystal experiments. In the 1990s, Emoto began photographing frozen water crystals under a microscope. He found that water exposed to different stimuli formed different crystal structures.

Water exposed to classical music, words like "love" and "gratitude," or prayers formed beautiful, symmetrical, snowflake-like crystals. Water exposed to heavy metal music, words like "hate" and "evil," or neglect formed chaotic, fragmented, irregular crystals.

He published his findings in a series of books, most famously, *The Hidden Messages in Water*, and the images went viral. People were captivated by the idea that water could "respond" to intention, emotion, and vibration. The scientific community, predictably, was skeptical. The experiments were not double-blind. The sample sizes were small. The selection of crystals photographed was subjective. The claims were extraordinary, and the evidence was not rigorous enough to meet scientific standards. And yet, the findings are hard to dismiss.

This is because we know that water is a highly structured, dynamic medium. It's not just H_2O molecules floating around randomly. Water forms hydrogen-bonded networks that can hold information and respond to electromagnetic fields. Water is influenced by vibration. This is not controversial. Ultrasound, for example, uses high-frequency vibrations to heat water, break up kidney stones, and image tissues inside the body.

Water is the primary medium of life. Your body is 60-70% water. Your cells are bathed in water. Your DNA is surrounded by structured water layers that influence how genes are expressed. If water can carry information, then the implications for biology are profound.

It is perceivable that Emoto's experiments were picking up something real, not proof of consciousness in water, but evidence that water's molecular structure is influenced by coherent vs. incoherent fields. Perhaps the "love" and "gratitude" experiments created coherence in the water, smooth, resonant, harmonious oscillations, and that coherence was reflected in the symmetry of the crystals. Or the "hate" and "neglect" experiments created incoherence, chaotic, dissonant, fragmented oscillations, and that incoherence was reflected in the disordered crystals. I'm not asking you to fully believe Emoto's claims without question. I'm asking you to consider the possibility that vibration and coherence matters, and that the quality of the field you're immersed in, loving or hostile, calm or chaotic, affects not just your mind, but your cells, your water, your very structure. Should that be true, then the frequency you resonate at is not trivial. It's foundational. (And it also lands us at the edge of epigenetics.)

The Biofield as Receiver and Transmitter

Let's come back to the biofield, the electromagnetic field generated by the body, particularly by the heart and the brain. We've already established that the heart generates a powerful electromagnetic field (60 times stronger than the brain's). This field extends up to 3 feet beyond the body. The pattern of the field changes based on emotional state, as coherent during love and chaotic during stress. Other people can detect this field, and their own biofields can entrain to it. Here's what we haven't said yet. The biofield is not just a transmitter. It's also a receiver.

Just as a radio antenna both sends and receives signals, your biofield both broadcasts your internal state and detects the fields around you. This is how you *sense* the presence of someone behind you. This is how you *feel* when someone is staring at you. This is how you *know* when someone is upset, even when they're smiling and saying they're fine. You're not reading body language (though that helps). You're not hearing their tone (which helps too). You're detecting their biofield, the electromagnetic signature of their nervous system, their heart, their state of being.

When that field is coherent, when the person is calm, present, and aligned, your biofield detects it and responds. You feel safe. You relax. Your own coherence increases. This is entrainment. This is resonance. This is love as a vibrational phenomenon.

The more coherent you become, the more sensitively you can detect these fields. This is why long-term meditators often report heightened sensitivity to the emotional states of others. This is why energy healers can "feel" blockages in a client's field. This is also why empaths are so affected by crowded spaces because they're picking up the electromagnetic noise of dozens or hundreds of incoherent biofields. Your biofield is a sensory organ. And what it senses is coherence, or the lack thereof.

If Everything Is Vibration, Love Is a Resonance We Attune To.

Here's the synthesis. Everything is vibration. Matter, energy, consciousness are all oscillating at different frequencies. Coherence is alignment. When oscillations synchronize, they amplify each other, creating order, health, and power. Entrainment is automatic. Oscillating systems naturally synchronize when in proximity. This applies to metronomes, fireflies, heart cells, and biofields.

Love is a coherent frequency in the field. It's not a single tone but a pattern, a resonance, a harmony, a state of alignment with the unified field, with Source, with the quantum vacuum. We detect love through our biofield. The heart, the cells, the nervous system are all are oscillators, capable of resonating with external fields. When we encounter coherence (in another person, in nature, in meditation, in the field itself), our system entrains. We synchronize. We align.

Traditional Love

So what about the person across the table? What about the one whose eyes you can't stop meeting? If love is a field we sense rather than an emotion we generate, then what happens when two people fall in love with each other? Here's what I believe is happening beyond the attraction, soul or karmic connection, or wound matching, each of which carries its own resonance. When you're in the presence of a healer or a teacher, their coherent biofield entrains yours. It's a one-directional gift. You receive. But when two people fall in love, something different occurs. Something mutual.

Your biofield reaches toward theirs and theirs reaches back. Your heart's electromagnetic field synchronizes with theirs and theirs amplifies in response. A feedback loop ignites. Your coherence deepens theirs. Theirs deepens yours. Back and forth, faster and faster, until something extraordinary happens and the two fields aren't just aligned. They're co-creating a shared coherence that neither could sustain alone. That shared coherence acts like a wider aperture.

Think of our hyperbolic slot. When you are alone and coherent, you are one plane intersecting the rod. The trace is real but modest. When two people enter mutual resonance, it's as if the opening in the plexiglass widens. The rod passes through more fully. The U-shape deepens. Source pours through with an intensity that takes your breath away.

This is why falling in love feels like remembering something ancient. It's not just the other person you're recognizing. It's the signal, suddenly amplified. (And of course the bonding neurochemicals come in to do their job!) This is why losing that person can feel like losing access to Source itself — because the aperture narrows and the trace becomes faint again. But the rod never moved. It was always there. The person was never the source of the love. They were a resonance partner, a co-creator of the opening through which love became undeniable. When you understand this, you stop clinging to people as the origin of what you felt. You honor them as the tuning fork that helped you widen your own capacity to sense what was always passing through.

This alignment is what we experience as love. Not the emotion (the body's reaction). Not the feeling (the mind's interpretation). But the sensing, the direct perception of coherence in the field.

Love is not something we create. Love is a frequency we tune into, and the more coherent we become through meditation, breathwork, coherence practices, energy work, or simply by being in the presence of coherent beings, the more clearly we sense it.

The Hierarchy of Response

Think of it as a spectrum of increasing sophistication in how different forms respond to the love frequency.

Water: Molecular organization shifts in response to the frequency. No choice, no awareness, just fundamental physical response to vibration.

Plants: Detection of loving versus harmful intent, measurable electrical responses, growth patterns that reflect the emotional environment. Still no choice in the conventional sense, but more complex responsiveness.

Animals: Bonding behaviors, attachment patterns, seeking of connection. Neurochemical amplification begins with oxytocin, dopamine, endorphins translating the frequency into behavior and biology.

Humans: All of the above, plus conscious awareness of love as a perceivable phenomenon.

We can detect it, amplify it through our biochemistry, *and* become conscious of ourselves as perceivers and participants in the frequency. We can cultivate our sensitivity to it, refine our perception, and intentionally work with it.

The frequency itself doesn't change across this spectrum. What changes is the sophistication of the *response system* — from water's molecular reorganization to human consciousness's ability to perceive, name, and intentionally cultivate love as a practice.

Returning to Avoidant Attachment

This framework illuminates why individuals with damaged oxytocin systems still seek love. They're not chasing a neurochemical reward their biology can't provide. They're responding to a frequency they perceive at the same fundamental level that plants and water perceive it, only they lack the *amplification system* that would translate that perception into the full-body experience of safety and bonding.

Their perception works. It's the amplification that's impaired. They're like plants detecting loving versus harmful intent, able to sense the frequency but unable to generate the complex biochemical cascade that creates the subjective experience of feeling loved. They know something real is there. Their bodies just can't properly translate it into the experience their human biology was designed to provide.

The Practical Implications

If love is vibration, then:

Your state matters. The frequency you're resonating at, either anxious or calm, scattered or coherent, reactive or present, affects not just you, but everyone around you. Your biofield is broadcasting. What are you transmitting?

You can train coherence. Just as you can train your muscles, your flexibility, your cardiovascular system, you can train your nervous system to hold coherent states. Heart coherence practices, such as meditation and breathwork are not "spiritual bypassing." They are nervous system training. They are biofield hygiene.

Environment matters. Just as plants grow better in coherent sound fields, you thrive in coherent environments. Chaotic, noisy, incoherent spaces drain you. Calm, natural, coherent spaces restore you. This is not psychological. This is biophysical.

You can entrain others and be entrained by them. When you're coherent, you help others become coherent. When you're incoherent, you contribute to the noise. This is not judgment. This is physics. And it's why being around certain people feels nourishing, while being around others feels exhausting.

Love is always accessible. The frequency is always there. The field is constant. But your ability to detect it depends on your coherence. When you're scattered, stressed, caught in emotional reactivity or mental loops, the signal gets drowned out by noise. When you're still, centered, present, the signal comes through clearly.

The Mystery Remains

Even with all this talk of frequency, coherence, and entrainment, we have not fully explained love. We have mapped it. We have described its behavior. We have shown that it follows the patterns of a vibrational phenomenon. But we have not answered the deepest question: What is it, ultimately?

Is love a property of the quantum field itself? Is it the "ground state" of the unified field, the natural resonance that emerges when nothing distorts it? Is love what consciousness feels like when it recognizes itself in another form? Is love the memory of unity, the trace left behind by our higher-dimensional self, intersecting this plane, reminding us of what we truly are?

I don't know. And neither does anyone else, not fully. But what I do know is that Love is not random. It's not arbitrary. It's not a neurochemical accident or an evolutionary glitch.

Love has structure. Love has coherence. Love has frequency. And you, with your heart, your biofield, your cells oscillating at gigahertz frequencies, are designed to sense it.

CHAPTER ELEVEN
The Mystical Evidence

Science can only take us so far. It can map the brain regions that light up during love. It can measure the electromagnetic field of the heart. It can detect quantum coherence in microtubules and biophotons in cells. It can show us that plants respond to intention and that biofields entrain.

But science cannot — not yet and perhaps not ever — fully explain what happens when you sit in the presence of a saint and feel the room fill with a palpable warmth. It cannot explain the tears that come when a Reiki practitioner's hands hover over your heart and something ancient and knotted suddenly releases. It cannot explain the phenomenon of remote viewing, where trained individuals accurately describe locations they have never seen, thousands of miles away.

Science can describe the effects. But it struggles with the cause. And that's because the cause is not material. It's not biochemical. It's not even electromagnetic, at least not in the conventional sense.

The cause is consciousness. Awareness. Presence. The field itself. This chapter is where we step fully into the mystical, not to abandon science, but to honor what science is beginning to approach from the edges. Because for thousands of years, mystics, healers, saints, and sages have been reporting the same thing — that love is not just an emotion. It is a transmission. A field. A presence that can be felt, shared, and known directly, without words, without touch, without any physical intermediary. They didn't have fMRI machines or magnetometers. They didn't publish peer-reviewed papers. But they knew. And now, science is beginning to catch up.

Reiki and the Transmission of Coherence

Let's start with something tangible. Reiki is a Japanese energy healing practice in which the practitioner channels life force energy (called ki or qi) through their hands, either by touch or by hovering the hands slightly above the body. The claim is that this energy promotes healing, balance, and relaxation.

From a conventional scientific perspective, this sounds absurd. How can someone's hands, not even touching you, create a physiological response? There's no massage. No medication. No physical mechanism. And yet, thousands of people including skeptics, doctors, nurses, and scientists report profound experiences during Reiki sessions.

- Deep relaxation, as if the nervous system suddenly downshifts
- Warmth or tingling in areas where the practitioner's hands are placed
- Emotional releases — tears, laughter, a sudden easing of long-held tension
- Pain reduction that lasts long after the session
- A sense of being "held" by something vast, benevolent, loving

Anecdotes, you might say. Placebo, perhaps. Wishful thinking. Yet it’s tangible. Reiki practitioners emit measurable magnetic fields from their hands during sessions. Studies conducted by researchers like Dr. John Zimmerman and Dr. James Oschman have detected pulsing biomagnetic fields emanating from the hands of healers — fields that are significantly stronger than the baseline magnetic field of the human body.

These fields oscillate at frequencies in the range of 0.3 to 30 Hz which are the same frequencies that have been shown to stimulate tissue repair, reduce inflammation, and promote healing. This is not imagination. This is detectable, repeatable, measurable.

Here's the deeper question. What is being transmitted? The conventional explanation would be electromagnetic energy. The hands generate a magnetic field, and that field influences the recipient's nervous system, heart rhythm, and cellular activity. None of this explains why Reiki recipients consistently report feeling love, not just relaxation or pain relief, but a specific, unmistakable sense of warmth, safety, and being cared for. This even happens when the practitioner is a stranger, when no words are exchanged, and when the touch is minimal or absent. What if Source coherence is being transmitted, not just energy?

What if the Reiki practitioner, through training, intention, and practice, has learned to bring their own biofield into a highly coherent state, and that coherence is what gets transmitted? (As a Reiki Master, I can attest to this practice.)

The recipient's nervous system, heart, and cells detect the coherence and entrain to it. Chaos becomes order. Dissonance becomes harmony. The body remembers what it feels like to be aligned. That alignment in the form of coherence, is what we experience as love. This would explain why:

- Reiki works even when the recipient is skeptical or unaware of what's happening.
- The effect is not just physical (pain relief) but also emotional (peace, safety, release).
- Different practitioners have different "qualities" of energy. Some feel warm, others cool; some feel gentle, others powerful. What's being transmitted is not a generic "energy," but the specific coherence of that practitioner's field to the greater Source field they are attuning to, acting as a channel for healing energy.

Reiki is not magic. It's resonance.

Darshan: The Presence of Enlightened Beings

Let's go deeper to darshan. Darshan is a Sanskrit word that means "sight" or "vision," but in the spiritual context, it refers to being in the presence of a holy person, a saint, guru, or enlightened being, and receiving a direct transmission of grace, blessing, or awakening simply by being near them. This is not a belief or faith. This is a phenomenon reported across cultures, religions, and spiritual traditions for millennia. Again, as with the same for Reiki and meditation, if I had not experienced it first-hand, multiple times with multiple beings, I would not have included it in the book as part of my experiential data to support this theory of love.

Whether it's:

- Ramana Maharshi sitting in silence on the mountain, and people from around the world traveling to sit near him and feel the dissolution of the separate self
- Amma, the "Hugging Saint," embracing millions of people and leaving them in tears, transformed
- Thich Nhat Hanh walking slowly through a meditation hall, and the entire room dropping into stillness
- Tibetan Monks laughing, and a stadium full of people suddenly feeling joy they can't explain

The experience is consistent. In the presence of certain beings, something shifts. The room becomes charged. The heart opens. The mind quiets. Love arises, not as an emotion directed toward the person, but as a field that fills the space. How do we explain this? From a materialist perspective, we can't. There's no chemical being transmitted. No words are being spoken (in many cases). No physical touch (most often). And yet, the effect is palpable, measurable in some cases (heart rate variability, brainwave patterns, cortisol levels), and life-changing for many.

What if these beings are not "more loving" in the emotional sense, but more coherent? Through decades of meditation, self-inquiry, and dissolution of ego, it is likely they have aligned their biofield so completely with Source, with the unified field, with the love frequency, that their mere presence acts as a tuning fork. When you sit near them, your biofield entrains. Your nervous system calms. Your heart rhythm becomes coherent. Your brainwaves synchronize. And suddenly, you're sensing the love frequency more clearly than you ever have before.

They are not giving you anything. They are simply being at a frequency so stable, so coherent, that it allows you to tune in and entrain your rhythm with theirs. This is transmission of coherence. Not of belief — coherence. And it's why darshan doesn't require understanding, language, or even spiritual context. Children feel it. Animals feel it. Skeptics feel it. That's because it's not psychological, it's vibrational.

Remote Viewing and Non-Local Consciousness

Next let's step into territory that makes even open-minded scientists uncomfortable: remote viewing.

Remote viewing is the practice of perceiving information about a distant or unseen target using extrasensory means. It was studied extensively by the U.S. military and intelligence agencies during the Cold War, in programs like Stargate Project at Stanford Research Institute (SRI). The results were startling. Under controlled, double-blind conditions, trained remote viewers were able to:

- Accurately describe buildings, landscapes, and objects they had never seen
- Identify hidden targets selected randomly by computer
- Provide actionable intelligence that was verified by later reconnaissance

This was not magic or guesswork. The statistical probability of getting these results by chance was astronomically low. The program ran for over 20 years, funded by taxpayer dollars, because it worked. When the program was declassified, many of the findings were published in peer-reviewed journals. Researchers like Dr. Hal Puthoff, Dr. Russell Targ, and Ingo Swann documented the phenomenon rigorously. Yet mainstream science still doesn't know what to do with it because remote viewing violates the fundamental assumption of materialism that states consciousness is local, confined to the brain, and cannot access information outside the reach of the physical senses.

But by now, many progressive people are beginning to come to terms with a model where consciousness is not local and not produced by the brain, but interfacing with the brain, tuning into a field of information that permeates all of space.

This is the hypothesis proposed by researchers studying non-local consciousness. The idea is that the brain is not a generator of consciousness, but a receiver, like a radio picking up a signal, or a lens focusing light. Were this true, then remote viewing makes sense. You're not "sending" your consciousness out to the distant location. You're tuning into the information field, sometimes called the Akashic field, the zero-point field, the unified field, where all information about all locations, past and present, is encoded.

Here's where it connects to love. If consciousness can access information non-locally, then it can also sense coherence non-locally. This is why you can *feel* when someone you love is in distress, even when they're far away. This is why mothers often *know* when their child is hurt. This is why people report sensing the moment a loved one dies, even across continents. You're not imagining it. You're sensing the field.

And love, as a coherent frequency in the field can be sensed non-locally, just like any other information encoded in the substrate of reality.

The Double-Slit Experiment and the Observer Effect

Let's briefly revisit one of the most famous experiments in quantum physics, the infamous double-slit experiment. Yes, we are going there because it's truly fundamental.

Here's the setup. You fire a stream of electrons (or photons) at a barrier with two slits. Behind the barrier is a screen that detects where the particles land. If you don't observe which slit the electron goes through, it behaves like a wave — passing through both slits simultaneously and creating an interference pattern on the screen (alternating bands of light and dark). If you do observe which slit the electron goes through, it behaves like a particle — going through one slit or the other, and the interference pattern disappears.

The act of observation changes the outcome, not because the measuring device physically disturbs the particle (this has been ruled out), but because consciousness, awareness, attention, the act of knowing, collapses the wave function. This is the observer effect, and it's one of the most incredible yet unsettling findings in all of physics. It suggests that consciousness is not a passive witness to reality, but an active participant. Reality does not exist independently of the observer.

Most physicists are careful to say this doesn't mean "consciousness creates reality" in a New Age sense. The observer effect is subtle, strange, and still not fully understood. Yet here's what's undeniable: Consciousness and matter interact at the quantum level. And if consciousness can influence the behavior of subatomic particles, then it's not a stretch to propose that consciousness can also influence biofields, cellular coherence, and the vibrational states of living systems.

This is the scientific foundation for energy healing, distant healing, prayer, and intention. It's not pseudoscience. It's quantum mechanics.

The Synthesis: What Mystics Always Knew

For thousands of years, mystics have been saying the same thing, in different languages, across different cultures.

Unity Consciousness says the separation between self and other is an illusion. At the deepest level, we are all one. *The Field* is represented too. There is a substrate of reality called Brahman, the Tao, the Dharmakaya, the Unified Field, God, Source, from which all things arise and to which all things return.

Love is the Ground of Being This substrate is not neutral. It is not cold. It is not indifferent. It is love. Pure, unconditional, boundless love. You don't need a priest, a book, or a doctrine to access this. You need only to turn inward, quiet the mind, open the heart, and sense what is already there. This is *Direct Knowing.*

For most of history, science dismissed this as poetry or wishful thinking, or perhaps as the projection of human emotion onto an indifferent cosmos. Yet now, quantum physics is saying the universe is not made of separate objects. It's a unified field, and consciousness is woven into the fabric of that field. Biofield science is saying the body is not an isolated system. It's a transmitter and receiver, resonating with fields around it. Quantum biology is saying that living systems use quantum coherence, entanglement, and non-locality. They are not classical machines. And finally, consciousness research is saying that awareness is not confined to the brain. It may be non-local, interfacing with a field of information that transcends space and time.

Science is beginning to map what mystics always knew. You don't have to choose. You don't have to be a scientist or a mystic. You can be both. You can honor the data and honor the direct knowing. You can measure the biofield and feel the transmission. You can study the brain and sit in darshan. They're pointing to the same thing. *Love is real. Love is a field. Love is perceivable. And you, with your biofield, your heart, your cells, your consciousness are naturally designed to sense it.*

The Evidence Is Not Just Scientific

The most compelling evidence for love as a sense is not in the lab. It's in lived experience. It's in the moment you sat in meditation and felt your chest open and warmth flood through you, even though there was no person, no memory, no reason for the internal biochemical cascade. It's in the time you walked into a room and immediately knew someone had just been crying, even though they were smiling. It's in the presence of that teacher, that healer, that friend whose very being makes you feel safe, seen, held. It's in the way your dog looks at you with such pure, uncomplicated love that you remember, just for a second, what it feels like to be whole. It's in the forest that hums with aliveness, the ocean that soothes something ancient in you, the sunrise that makes you cry for no reason you can name.

You know love when you sense it. And the fact that you know, without needing proof, without needing explanation, is itself the evidence. Science can support this. Science can measure the biofield, track the brainwaves, detect the coherence. But science cannot replace the direct experience. And it shouldn't.

You are the instrument. Your body, your heart, your awareness are the tools by which love is known. When science and mysticism finally converge, when the data and the direct knowing point to the same truth, then we will no longer need to defend, justify, or prove what we sense. We will simply recognize it, trust it, and follow it home.

The Invitation

Here is the invitation this chapter offers. Stop waiting for science to validate what you already know. If you have felt love arise in meditation, trust it. If you have sensed the presence of a coherent being, honor it. If you have received healing through touch, energy, or prayer, accept it. If you have known things you couldn't possibly know through logic, believe it. You are sensing the field and the field is real. Science is catching up. Slowly, carefully, with the appropriate skepticism and rigor. And that's good. We need the science. We need the maps, the models, the measurements.

We also need the mystics. We need the lived experience. We need the people who have touched the field and returned to say "Yes. It's real. And it's made of love." Your direct knowing and your sensing of the love frequency is not less valid than a peer-reviewed study. It's the data the studies are trying to explain. You are the evidence.

CHAPTER TWELVE
Love as Evolutionary Compass

We've established that love can be classified as a sense. We've shown that it meets all the criteria because it has a stimulus (coherence in the field), receptors (D2/D3, interoceptive pathways, biofield sensitivity), transduction (biochemical cascade), neural pathways (limbic system to cortex), and conscious perception (the unmistakable felt experience). We've explored the biology, the quantum mechanics, the vibrational nature, the mystical evidence. Next we need to ask the evolutionary question of *why*?

Why would evolution design us this way? Could natural selection favor a sensory system that detects something as seemingly "impractical" as love, especially unconditional love, which offers no survival advantage, no reproductive benefit, no immediate resource gain? If evolution is about survival of the fittest, why would we evolve a sense that makes us yearn for connection with all beings?

Why would we evolve a nervous system that lights up in the presence of coherence, even when there's no food, no mate, no threat to defend against?

The materialist answer is, "*we didn't*". Love is a byproduct, a side effect, a spandrel — an evolutionary architecture that exists not because it was selected for, but because it piggybacked on other traits that were useful. But I don't think that's the full story.

I think love is not a byproduct of anything. I believe it's a compass and a navigational system — a homing beacon. It's an evolutionary adaptation designed not just to help us survive in three-dimensional reality, but to guide us beyond it, back to the multidimensional awareness, the unity consciousness, the Source from which we came. This chapter explores that possibility. And if it's true, then love is not just another sense. It's the sense that could define humanity's next evolutionary leap.

The Paradox of Unconditional Love

Let's start with the problem. From an evolutionary perspective, attachment makes sense. Bonding with a mate increases the likelihood that offspring will survive. Bonding with kin ensures genetic continuity. Even bonding with a tribe improves chances of survival through cooperation, resource sharing, and collective defense. This is why dopamine, oxytocin, and vasopressin evolved. They reinforce behaviors that are advantageous for survival and reproduction. You feel good when you bond with someone because bonding increases fitness.

Unconditional love doesn't fit this model. Unconditional love arises in meditation, when there is no object, no relationship, no evolutionary advantage. It arises in the presence of strangers, animals, even inanimate nature. It can arise spontaneously, without cause, without reason. It doesn't make you better at hunting. It doesn't make you more fertile. It doesn't increase your social status or your access to resources. In fact, it often makes you less competitive, more open, more vulnerable, more willing to sacrifice your own interests for the well-being of others.

From a pure survival standpoint, unconditional love is a liability. Yet it persists across cultures, across centuries and across individuals who have no genetic relationship, no shared environment, no common conditioning. Why? The conventional answer is that unconditional love is a cultural phenomenon and a moral ideal taught by religions, a virtue cultivated by spiritual practice, a psychological state achieved through self-development. But that doesn't explain why it feels so fundamental, why it arises unbidden in meditation or why children, before they've been taught any moral framework, can exhibit profound compassion. Or why even in the absence of cultural reinforcement, people report the same experience of a boundless, objectless love that feels more real, more true, more natural than any conditioned emotion.

What if the reason love doesn't serve survival is because it's not meant to? What if it's meant to serve something else — something evolution has been guiding us toward all along?

Evolution Is Not Just About Survival

This is where we get to update our understanding of evolution. For most of the 20th century, evolutionary biology was dominated by the gene-centered view. Organisms are vehicles for genes, and natural selection favors genes that maximize reproductive success. Survival and reproduction are the only currencies that matter.

This view is incomplete. Evolution is not just about survival. It's also about complexity, integration, and consciousness. Let's look at the trajectory of life on Earth.

Single-celled organisms → multicellular organisms
Simple nervous systems → complex brains
Reflexive behavior → learning, memory, and planning
Solitary individuals → social groups
Instinct-driven animals → self-aware beings capable of language, art, and abstract thought

At every stage, evolution has moved toward greater complexity, greater integration, and greater awareness. Once a certain threshold of complexity is reached, organisms begin to exhibit behaviors that transcend survival. They create art. They explore the cosmos. They contemplate their own existence. They seek meaning, beauty, truth, and things that have no direct survival value, but seem to be inherent to the structure of consciousness itself.

Love must be part of that trajectory. What if the evolution of the love sense is not about pair bonding or kin selection, but about the expansion of awareness itself? Perhaps love evolved not to make us better survivors, but to make us better perceivers, capable of sensing and aligning with the multidimensional coherence from which all life emerges. Perhaps love is taking us home, to seek relentlessly for what we really are.

The Function of Love: Remembering

Let me propose a new framework for understanding the function of love.

Love is not about bonding with others. It's about remembering the Self.

I'm not talking about the small self, the ego, the identity, the story you tell about who you are, but the Self with a capital S — the unified field of consciousness, the Source, the infinite awareness of which you are a localized expression.

From this perspective, the reason we search for love so relentlessly is not because we lack it, but because we remember it. On some level, cellular, quantum, dimensional — we know what we are. We know we are not separate. We know we are held, loved, and inseparable from the field but in this body, in this density, in this third-dimensional reality, we forget.

The boundaries of the body create the illusion of separation. The survival instincts create the sense of self vs. other. The noise of the mind drowns out the signal. So we long for more. We search, endlessly. We reach for connection, intimacy, belonging, anything that can remind us, even briefly, of what we once knew. The longing for love is the longing for unity. And unity is not something we have to create. It's something we have to remember.

Romantic love is not about finding "the one." It's about finding someone whose presence reminds you of the field, allows you to sense the frequency more clearly. Unconditional love is not something you achieve. It's what arises when the noise quiets and you sense the field directly. Spiritual practice is not about becoming enlightened. It's about removing the obstructions so you can sense what was always there.

Love is the compass that points us back to Source. And we evolved the ability to sense it precisely because Source is what we are.

Why D3 Receptors Evolved This Way

Let's come back to the biology for a moment. The D3 dopamine receptor is unique. It's highly localized in the limbic system, the emotional, reward, and memory centers of the brain. It's more sensitive than other dopamine receptors, responding to lower concentrations of dopamine. It's involved in sustained states of reward and well-being, not just fleeting pleasure.

From a survival perspective, D3 doesn't make much sense. D1 and D2 receptors are sufficient for motivation, movement, and reward-seeking behavior. Why would evolution create a third receptor, so specific, so sensitive, so tied to deep, objectless states of bliss? It's not about survival. It's about sensing the field.

What if D3 receptors evolved as the biological interface for detecting coherence? What if they're tuned to the love frequency — *a resonance of the unified field* — and their activation is the body's way of signaling, "You're aligned. You're home. This is what you've been searching for."

This would explain why D3 activation creates a sense of "rightness", not just pleasure, but meaning. D3 is tied to transcendent states and mystical experiences, flow states, deep meditation. D3 dysfunction is linked to anhedonia (inability to feel pleasure) and existential disconnection, the sense that life has no meaning, no purpose, no coherence.

D3 receptors are potentially the body's love sensor. And just maybe, they evolved to guide us toward coherence, not just toward survival.

Neuroplasticity and the Expansion of Perception

Our brain is not fixed. For most of the 20th century, neuroscience believed that the adult brain was static; you were born with a certain number of neurons, and once they were formed, that was it. You couldn't grow new ones. You couldn't change the wiring.

In the 1990s, researchers discovered neuroplasticity which is the brain's ability to reorganize itself, form new connections, and even grow new neurons throughout life. This was revolutionary. It meant that experience shapes the brain. Meditation changes the brain. Learning changes the brain. Trauma changes the brain. And healing changes the brain. Perception itself can change the brain.

When you train your attention through meditation, through sensory refinement, through any practice that sharpens awareness, you literally rewire the neural pathways involved in that perception. Musicians who train their hearing develop denser auditory cortices. Meditators who focus on the breath develop stronger interoceptive awareness. Sommeliers who refine their sense of taste develop more nuanced gustatory processing. The more you use a sense, the sharper it becomes. And the sharper it becomes, the more of reality you can perceive.

Here's the ultimate question. What happens when you train the love sense? What happens when you meditate regularly, practice heart coherence, cultivate presence, and intentionally attune to the frequency of love? You don't just "feel better"; you change your brain. You strengthen the neural pathways involved in detecting coherence. You refine your biofield sensitivity. You become more capable of sensing the love frequency, not just in peak moments, but in everyday life. This is neuroplasticity applied to the sixth sense. If enough people do this, if humanity as a whole begins to train this faculty, to recognize it as a sense, to refine it intentionally, then we could be on the edge of an evolutionary leap.

The Next Stage of Human Evolution

Evolution doesn't stop. It's not done with us. We are still evolving, perhaps not just biologically (which happens too slowly for us to notice), but neurologically and consciously.

The human brain has tripled in size over the past two million years. Language emerged. Abstract thought emerged. Self-awareness emerged. And each of these capacities expanded what we could perceive, understand, and create. What if the next or one of the next stages of human evolution is the conscious refinement of the love sense? What if the trajectory of human development is moving us from survival-based consciousness (fear, scarcity, competition) → relational consciousness (empathy, cooperation, compassion) → unity consciousness (direct perception of the unified field, love as the ground of being)? This is the direction mystics, sages, and spiritual traditions have been pointing toward for millennia. Now, science is beginning to provide the framework:

Quantum biology shows us that life operates at the quantum level, using coherence, entanglement, and non-locality. Neuroscience shows us that the brain is plastic, capable of expanding its perceptual capacities. Biofield science shows us that we are not isolated systems, but transmitters and receivers resonating with the larger field. Consciousness research shows us that awareness is not confined to the brain, but may be non-local, interfacing with a unified field of information. All of this points toward the same conclusion. *We are designed to sense the field.* And the more we refine that sense, the more we align with what we truly are.

This is not evolution through genetic mutation. This is evolution through conscious refinement of perception. It's already happening. Millions of people are meditating, practicing mindfulness, exploring plant medicine, doing energy work, seeking direct experience of the field. They're not doing it because a scientist told them to. They're doing it because something in them knows that there's more, knows that the boundary between self and other is not as solid as it seems, knows that love is not just an emotion but a dimension of reality. They are training the love sense. And in doing so, they are evolving.

Love as the Compass Home

Evolution did not give us the love sense so we could bond with a mate. It gave us the love sense so we could find our way back to Source.

The longing for love is not a flaw. It's not neediness, weakness, or attachment. It's a *navigational instinct* — the body-mind's way of orienting itself toward the frequency of unity, coherence, and wholeness.

When you feel love whether in meditation, in nature, in the presence of another being, or in the silence of your own heart, you are not creating it. You are sensing it. You are tuning into the field that was always there, the ground of being that holds all things together.

The more you train this sense, the more clearly you can sense it. The more clearly you sense it, the more you align with it. And the more you align with it, the more you remember what you are. You are not a separate self, isolated in a body, lost in a cold universe. You are a localized expression of the unified field, temporarily condensed into form, sensing your way back to the wholeness you never truly left. Love is the compass. It always has been.

Perhaps for the first time we have the language, the science, and the collective readiness to recognize it for what it is. Love is not a feeling or emotion. *Love is a sense — the sense that could guide humanity home.*

CHAPTER THIRTEEN
The Next Human

So now you know. You are armed with all the information to provide a new option to see the world differently. In my world, and maybe yours now too, love is not an emotion you produce. It's a frequency you detect. It's a sense, with receptors, pathways, transduction mechanisms, and conscious perception. It's vibrational. It's quantum.

It's real. Plants sense it. Animals sense it. Your heart senses it. Your cells sense it. And you, the conscious, aware, multidimensional being reading these words, have always been sensing it, even when you didn't have the language to name it.

The longing you feel for love is not a flaw. It's a compass and a homing signal and an evolutionary adaptation guiding you back to the unified field, to Source, to the remembering of what you truly are.

You know all of this now. So the question becomes, what do you do with this knowing? How do you live from it? How do you train the love sense, the way you would train your ear for music or your palate for wine? How do you move through the world as someone who is not searching for love, but sensing it, tuning into it, aligning with it, transmitting it? What does it mean to be the next human? This chapter is the bridge between understanding and embodiment.

What Changes When You Treat Love as a Sense?

Let's start with the shift in perspective. If love is an emotion, then you're at the mercy of circumstance. You feel love when the right conditions are met such as when someone treats you well, when life goes your way, when the chemistry is right. And when those conditions change, the love fades. You fall in and out of it, helplessly, like a leaf blown by the wind.

If love is a sense, then you are not at the mercy of circumstance. You are a perceiver. Just as you can train your eyes to see more clearly, your ears to hear more subtly, your body to move with greater precision, you can train your ability to sense love. You can refine the instrument by clearing the noise. You can attune to the frequency, regardless of what's happening around you. I don't mean by spiritual bypassing, toxic positivity, or pretending everything is fine when it's not.

This is recognizing that love, the coherence in the field, the resonance with Source, is always present, even in the midst of pain, grief, chaos, or loss. It's not that the pain isn't real. It's that the love is also real. When you can sense both simultaneously, you are no longer overwhelmed by one or the other. You are held by something larger.

What Changes With The Love Sense

1. You stop seeking love solely from others and start sensing it in the field.

This doesn't mean you stop wanting connection, intimacy, or partnership. It means you stop making other people responsible for giving you what was never theirs to give.

You recognize that the love you feel in the presence of another person is not coming from them. It's the coherence of their biofield allowing you to tune into the love frequency more clearly. They are a tuning fork, not the source of the sound.

When you understand this, relationships become lighter. There's less grasping, less fear of loss, less neediness. Because you know that even if the person leaves, the love doesn't. The field remains.

2. You stop confusing emotional reactivity with love.

You begin to distinguish between *emotion*, which is the rush, the intensity, the drama, the biochemical cascade triggered by attraction, attachment, or fear of loss, and *feeling*, which is the story you tell yourself about what the emotion means, such as, "I can't live without them," "They complete me," "I'm not lovable unless they love me back."

You come into *sensing* which is the direct, quiet, unshakable perception of coherence in the field, present whether or not the emotion or feeling is there.

This doesn't mean emotions are bad or wrong. Emotions are information. They're part of being human, but they are not love itself. And when you can sense love beneath the emotions like the ocean beneath the waves, you are no longer tossed around by every surge and crash.

3. You become more discerning about who and what you let into your field.

When you understand that your biofield is a sensory organ, you start paying attention to what affects it. Certain people, places, and environments enhance your coherence. They feel nourishing, expansive, clarifying. Others disrupt your coherence. They feel draining, chaotic, dissonant. This is not judgment. This is biofield hygiene. Just as you wouldn't drink contaminated water or breathe polluted air, you learn not to immerse yourself in incoherent fields unless absolutely necessary, and when you must, you learn how to maintain your own coherence so you're not destabilized.

4. You start transmitting coherence, not just receiving it.

The more you refine your ability to sense love, the more coherent you become. And the more coherent you become, the more you transmit that coherence to others. You become the tuning fork. You become the Reiki practitioner. You become the enlightened being whose mere presence allows others to sense the field more clearly. This is the beautiful paradox of love. It is not about becoming special or superior. It's about recognizing that coherence is contagious. When you are calm, others calm. When you are present, others become present. When your heart is open, others feel safe to open theirs. You are not just sensing the love frequency. You are becoming it.

How to Train the Love Sense

How do you refine the love sense, clear the noise, and attune to the frequency? Here are a few practices that are simple, accessible, and grounded in both ancient wisdom and modern science.

Practice 1: Heart Coherence Breathing

The heart is your primary love sensor and when the heart is coherent, meaning when its rhythm is smooth, rhythmic, and harmonious, the entire nervous system shifts into a state of receptivity.

The Process:

Sit comfortably. Close your eyes. Place one hand on your heart. Breathe slowly. Inhale for a count of 5. Exhale for a count of 5. Keep the breath smooth, even, natural — no forcing.

Focus on the heart. As you breathe, imagine the breath flowing in and out of your heart center (the middle of your chest). Feel the rise and fall with each breath.

Cultivate a feeling of appreciation. Think of something or someone you're grateful for — a person, a pet, a place, a moment. Let the warmth of that gratitude fill your chest.

Rest in the coherence. Stay with the breath, the heart, and the feeling for 5-10 minutes. Notice how your body softens. Notice how the noise in your mind quiets. Notice the sense of being held.

This practice shifts your heart rhythm into a coherent state, which synchronizes your brain, calms your nervous system, and opens your biofield to receive the love frequency. Studies have shown that just 5 minutes of heart coherence breathing can measurably improve heart rate variability, reduce cortisol, and increase feelings of well-being.

Do this daily. Morning or evening, or both. Think of it as tuning your instrument before you play.

Practice 2: Meditation on Unconditional Love

This practice is to bypass the emotional loop entirely and go straight to the sensing. It trains your nervous system to sense love without an object, without a trigger, without needing anything external. It reveals that love is not something you create, but something you perceive when the noise clears.

The Process:

Sit in stillness. No music, no guidance, no distractions — just you and the silence. Drop the object, meaning don't focus on a person you love.

Don't visualize anything. Don't try to generate a feeling. Just sit. Notice what arises. As the mind quiets and the body settles, pay attention to only the space between your eyes. Return there when the mind strays.

Notice if there's warmth, expansion, softness in the body, once the mind is quiet. Notice if there's a sense of being held, even when you're alone.

Don't grasp it. If love arises, don't try to hold onto it. Don't analyze it. Don't think, "Oh good, I'm doing it right." Just let it be. Let it move through you like a wave.

Return when you drift. When the mind wanders (and it will), gently return to the body, the breath, the heart. Over and over. No judgment. Just return.

Do this regularly. Even 10-20 minutes a day will, over time, rewire your brain and refine your sensitivity to the field.

Practice 3: Somatic Tracking

This is about developing interoceptive awareness — learning to sense the subtle signals in your body that indicate you're tuning into coherence (or losing it).

The Process:

Throughout the day, pause and ask yourself:

- Where do I feel this (feeling/emotion/thing) in my body? (Not "What am I thinking about this?" but "Where is this landing in my soma?")

- Does this feel expanding or contracting? (Coherence feels open, spacious, soft. Incoherence feels tight, heavy, closed.)
- Am I sensing love, or am I reacting to its absence? (There's a difference between the ache of longing and the direct sensing of presence.)

If you experience tension, breathe into that spot to release. If you are feeling peace, love, let that naturally expand with each breath.

This practice trains you to use your body as a biofeedback instrument. Over time, you become so attuned to the signals that you can tell, in real time, whether you're aligned with coherence or caught in reactivity.

Do this throughout the day. Especially in challenging moments. It's a form of mindfulness, but grounded in the soma, not the mind.

Practice 4: Coherent Relating

This is about bringing the love sense into relationship, not as a concept, but as a practice.

The Process:

Sit facing another person. A partner, a friend, anyone willing to try this with you.

Breathe together. Synchronize your breath for a few minutes. Just breathing, in silence, maintaining eye contact if that feels comfortable.

Sense, don't think. Don't try to "send love" or "receive love." Just sense what's present in the space between you. Is there warmth? Resonance? A palpable field?

Notice any possible entrainment. Pay attention to how your nervous systems affect each other. Does your heart rate slow? Does your breath deepen? Do you feel more grounded, more present, more connected?

This practice reveals that love is not inside you or inside the other person — it's in the field between you. When two nervous systems become coherent together, the love frequency amplifies. You're not creating it. You're tuning into it, together.

Do this with intention. Not every day, but regularly enough that it becomes a touchstone for what coherent relating feels like.

Practice 5: Biofield Hygiene

Just as you brush your teeth, shower, and eat healthy food to maintain physical health, you need practices to maintain biofield coherence.

The Process:

Spend time in nature. Trees, water, mountains, open sky are all are coherent fields. They don't have the electromagnetic noise of cities, the emotional chaos of crowds. They remind your system what coherence feels like.

Limit exposure to incoherent environments. If you must be in a crowded, noisy, chaotic space, practice grounding techniques: breathe, feel your feet on the ground, imagine roots extending into the earth. Don't absorb the chaos; let it move around you.

Clear your space. Your home, your workspace all affect your biofield. Clutter creates mental and energetic noise. Clean, simple, intentional spaces support coherence.

Choose your media carefully. What you watch, read, and listen to affects your nervous system. If it's agitating, fear-based, or incoherent, it will disrupt your ability to sense love clearly.

Your biofield is constantly interacting with the fields around you. The more you protect and nourish it, the more sensitively you can sense the love frequency. Do this as lifestyle. Not as a rigid rule, but as a gentle attunement to what supports your coherence and what doesn't.

What It Looks Like to Live This Way

When you begin to treat love as a sense, and when you train it, refine it, and live from it, your life changes in subtle but profound ways. You become less reactive because you're no longer at the mercy of emotional surges. You can feel the emotion, acknowledge it, and still sense the love beneath it.

You become more present because presence is coherence and coherence is what allows you to sense the field. You become less attached to outcomes because you're not trying to get love from people or circumstances. You're tuning into it, directly, in the moment. You become more discerning, not judgmental, but clear. You know what nourishes you and what depletes you and you honor that knowing.

You become a stabilizing presence. Not because you're trying to be, but because coherence naturally radiates. People feel calmer around you. Animals trust you. Spaces you enter feel different.

You stop searching and start sensing. The relentless hunt for love in relationships, achievements, substances, experiences softens because you realize it was never out there. It's here. It's always been here. You just needed to remember how to perceive it.

The Invitation

This is not a book that ends with a conclusion. This is a book that ends with an invitation.

I invite you to stop waiting for someone to love you into wholeness.

I invite you to stop believing that love is scarce, conditional, or something you have to earn.

I invite you to recognize that you are a sensory instrument, tuned to a frequency most of humanity has forgotten how to hear.

I invite you to train that sense through meditation, breath, presence, and practice, until the love frequency becomes as clear to you as the sound of your own heartbeat.

I invite you to live as a coherent being, not perfect, not enlightened, not beyond pain or struggle, but aligned, attuned, and awake to the field.

This is what the next human looks like. It's not a superhuman and not transcendent in the way we've imagined it. It's a human that is simply more aware, more sensitive and more capable of sensing what was always there.

You don't need special powers. You don't need to sit in a cave for 40 years. You don't need a guru, a teacher, or a certificate. You just need to tune in. To breathe. To sense. To open. To trust that when the noise quiets, love is what remains because love is not a reward for good behavior. It's not a gift bestowed upon the worthy. It's not a fleeting state you fall into and out of by accident.

Love is the ground of being. It's the substrate of reality. It's the frequency of Source itself. And you with your D3 receptors, your heart's electromagnetic field, your interoceptive pathways, your biofield resonating at gigahertz frequencies are designed to sense it.

So sense it. Train the faculty. Clear the static. Tune the instrument. And when you do, you won't just be perceiving love. You'll be living it. Transmitting it. You'll be the love frequency made flesh. And that — that is the next human.

EPILOGUE
The Field Remembers

For the entire journey of this book, we've been looking at the trace, the U-shaped cutout on the plexiglass plane — the shadow cast by something higher-dimensional as it passes through our limited perception. We've examined the clues, the patterns, the signals that tell us something vast is here, even when we can't see it fully.

We've followed those clues through biology, quantum physics, plant intelligence, mystical transmission, and evolutionary theory. We've built the case, step by step, that love is not an emotion but more likely a sense — a faculty by which we perceive the coherence of the unified field, the frequency of Source itself. But all of this, the science, the metaphors, the practices, can only take us so far. Because at some point, the question is not "Is this true?" but "Have you felt it?" Have you touched the field directly? Have you sensed the love frequency so clearly, so completely, that there was no room left for doubt?

I have. And I want to tell you about it. It's not proof or another argument. It's my testimony as a witness. As perhaps too, it's an invitation for you to trust what you, too, have sensed or will sense, when the time is right. This is what happened when I stopped looking at the trace and saw the rod.

South India, Late 2016

It was near the end of an idyllic 30 days on retreat at a spiritual campus in South India. The small number of students who signed up for this immersion were lucky — the program was almost cancelled right at the time I was about to board the plane, and the campus staff were attempting to push us all into a week-long program at more than three times the cost and with tenfold the number of people. Grace intervened, and I was grateful to be there with only a couple of dozen fellow travelers from around the world instead of the hundreds that attended most of the campus programs.

Every morning we awoke to pinkish-purple pre-dawn skies with the crowing of roosters, and soon after, spiritual hymns and chants wafted into the dorms from somewhere far away. We spent the days in meditation, yoga, contemplation, and bliss, and let the magic energy of India's heart melt ours.

At least once a week we would attend a spiritual process or darshan (energy blessing) at the massive and serene marble temple, and this one would be my last before leaving the quietude of this around-the-clock spiritual immersion.

In the golden rays of the setting sun, we filed up the temple stairs and introspectively walked clockwise around the second floor of the building multiple times before entering. Not too far across the floor, groups of monks were engaged in meditation and puja and some kind of ceremony with flowers that I couldn't identify. Our group was guided to sit cross-legged on the cold stone floors in a corner of the massive room, and once there, we breathed in golden light and waves of peaceful bliss. Sanskrit mantras were chanted softly in the background, and darshan began. I'd sat in the presence of multiple gurus before, but this experience overshadowed anything I thought I knew, far beyond any prior darshans.

In the presence of no human bodied guru, soon the powerful energies washed across the stillness in the open space — so strong that they pulled me out of peaceful meditation. Every cell of my body became activated and vibrating in a way I had never felt before.

The energy escalated to a fever pitch, and I was filled with the most powerful and heightened state I had ever experienced. Peace and bliss were overshadowed by my heart and solar plexus being so filled with love and abundance and divine grace that I could only weep in gratitude. Soon 'I' was gone, and an immense, energetic presence took over completely.

The Dissolution

In its place was everything and nothing. The only experience that remained happened in waves and waves of overwhelming full-body ecstasy, bliss, and love — peace and abundance that rippled out and in, to and from every part of the ether beyond the space of this body, the temple, and onwards.

Tears were streaming. The heart poured open. And a radiance like I have never felt before or after filled every essence of my being, merging without borders or boundaries into the infinite. I was dispersed into a cloud of bliss and love.

Every single one of my desires felt fulfilled — mentally, emotionally, and energetically — and I was full in a manner of overflowing abundance that I have yet to experience again. The feeling of tremendous, exuberant, and massive bliss, vibrancy, wealth, and riches — not just financially but in everything that existed — became a part of me. I became love, overflowing.

I merged into what I experienced as existence itself, and understood that this indescribable magnitude of non-stop abundance and gifts is the state that exists right at the other side of our mental and energetic separation from it.

Gratitude and love were the only senses of self that was left, as this ultimate peak experience beyond thought and emotion took hold quite naturally. In the disappearance of the I, there was only vibration. Only sensation. Only the fulfillment of every positive thought and desire and wish I had ever had. And no words to explain any of it.

What I Was Sensing

Deep in this darshan, with no human guru present, the force of our Universe heightened my state of consciousness to gift me the energetic experience of complete fulfillment and gratitude so it would become something my energy body could tangibly know, and then repeat. This was not imagination. This was not projection. This was not even a feeling in the conventional sense. This was direct perception.

Every concept we've explored in this book — the biofield, the quantum coherence, the heart's electromagnetic resonance, the D3 receptors lighting up, the interoceptive pathways carrying signals from the body to the brain — all of it was happening simultaneously, at a level of intensity I had never experienced before. I was sensing the field. This was a full-bodied experience and my own empirical evidence, not an abstract idea. It wasn't a hope or a belief. It was a living, palpable, undeniable presence, a frequency so coherent, so vast, so saturated with love that it dissolved every boundary between self and other, inside and outside, me and the infinite.

This was the love frequency. And for those moments — minutes? hours? time had no meaning. I was it. I didn't sense it from a distance. Nor did I receive it from somewhere else. I was being it. The instrument and the signal, inseparable.

The Return

When I returned to the United States over a month later, many things in my life changed because of this experience. I began to receive an endless stream of free or incredibly reduced-cost little items

I had at some time in the past thought about acquiring or imagined would be nice to have. Synchronicities multiplied. Doors opened. The universe seemed to conspire in my favor in ways that felt almost absurd.

Yet more than the external shifts, something internal had changed. I had been given a map, an energetic memory, and a felt knowing of what the substrate of existence is made of. It wasn't neutral. It wasn't cold. It wasn't indifferent. It was love. (or how we experience it in our human bodies) Overwhelming, abundant, boundless love, woven into the fabric of reality itself.

This is what the mystics mean when they say "God is love." This is what the quantum physicists are pointing toward when they describe the unified field. This is what every spiritual tradition has been trying to articulate in its own language for millennia. The ground of being is not empty. It is full. Overflowing. Radiant. Alive. And with our biofields, our hearts, and our cells oscillating at gigahertz frequencies, we are designed to sense it.

Why I'm Telling You This

I'm not sharing this story to convince you that my experience was unique. I'm sharing it because it's *not* unique. Thousands of people have had experiences like this in temples, in meditation halls, in nature, in near-death experiences, and in moments of profound grief or joy. The details differ, but the essence is the same. The boundary dissolves. The separate self falls away. And what remains is love — pure, objectless, all-encompassing love. This is not mystical hyperbole. This is direct perception of the field.

You don't need to go to India and you don't need a guru. You don't need a marble temple or a month-long immersion. The field is here. Now. Always. You are already in it. You are already made of it. You just need to remember how to sense it.

The practices in Chapter 12 are not about creating something new. They're about clearing the noise so you can perceive what was always there and tuning the instrument so the signal comes through clearly.

When it does, even for a moment in the midst of an ordinary day, you will know. You will know that the longing you've carried your whole life was not a sign of lack, but a homing signal. You will know that the love you've been searching for was never in another person, another place, another circumstance. You will know that it was here all along, waiting for you to tune in.

The Rod Was Always There

Remember the rotating rod and the plexiglass plane? For most of your life, you've been looking at the U-shaped trace — the shadow, the fragment, the clue that something larger is intersecting your reality. You've been trying to understand love by studying the trace including the emotions it triggers, the relationships it arises in, and the moments it visits and then fades. *The trace is not the rod. The shadow is not the object.* The rod is always there, rotating above the plane, whole and coherent and real, even when you can't see it fully.

Love is always there. The field is always there. The coherence, the resonance, the frequency of Source never leaves. You are the one who moves in and out of alignment with it.

The love sense our real sixth sense, our forgotten faculty, is what allows us to detect it, tune into it, and remember what we are.

Something Important to Remember

Trust what you have sensed.

Trust the moment your heart opened for no reason.

Trust the peace that arose in meditation when there was no object, no person, no cause.

Trust the presence you felt in the forest, by the ocean, under the stars.

Trust the love that flooded through you when you held your child, your pet, your beloved — not because of who they are, but because their coherence allowed you to tune into the field more clearly.

Trust the ache, the longing, the relentless search, not as evidence that you are missing something, but as evidence that you are sensing something real.

You are perceiving the field. And the field is love.

Tune in. Clear the noise. Refine the instrument. Practice the coherence. When the love arises — and it will, again and again, in ways you cannot predict or control — don't grasp it. Don't make it mean something about you or your worthiness. Just sense it. Receive it. Let it move through you because that's what you're designed to do. You are not here to create love. You are here to sense it, transmit it, and remember what you are. You are the love frequency made flesh. And the field — vast, coherent, infinite — has been waiting for you to remember.

Full Circle

We began this book with a question: What if love is a sense? We followed the clues through biology, quantum physics, plant intelligence, and mystical transmission. We built the case, step by step, that love is not an emotion but a dimensional signal, a frequency we detect through the biofield, the heart, the cells, the nervous system. And now we close with the only answer that matters.

Yes. Love is a sense.

You have always been sensing it. The rod was always there, rotating through your reality, leaving traces you could feel but not fully see. With the framework, the practices, the permission to trust your direct knowing, you can sense it more clearly. Love is not something you achieve. It's something you attune to and something you remember. *Welcome home to your true nature of love.*

Notes on Sources & Evidence

This book is not written as a scientific treatise or a proof, but as a working hypothesis — a model that invites us to reconsider love not only as an emotion, but as a perceptual faculty, a sense, and possibly a form of interconnection woven into the fabric of reality itself.

Because of this, the evidence referenced throughout the text falls into three distinct categories. To honor clarity and intellectual honesty, and to respect both scientific rigor and experiential knowing, I offer the following guide to how information is used here.

1. Established Scientific Evidence

Findings widely recognized and replicated within scientific literature.

These include:

Neuroscience of love and attachment

Oxytocin, dopamine, vasopressin, and bonding biology

Reward circuitry activation during romantic love (fMRI)

Heart rate variability and psychophysiological coherence

Interoception as a sensory system

Meditation-associated brainwave and neurochemical changes

Evolutionary links between social connection and survival

This category represents the empirical backbone of the book, the areas where love, connection, physiology, and perception already have known, measurable correlates. These findings are cited through peer-reviewed publications, meta-analyses, and established scientific institutes.

2. Emerging or Frontier Research

Scientifically investigated, but not yet universally accepted or fully understood. This includes studies involving:

Coherence and synchrony between individuals

Possible human magnetoreception

Bioelectric and electromagnetic signaling in heart-brain interaction

Quantum coherence in biological systems

528 Hz frequency, cortisol/oxytocin implications

Physiological entrainment between humans, plants, and animals

These fields sit at the boundary between known and unexplored, where science observes real phenomena but does not yet fully explain them. They are neither dismissed as fantasy nor elevated to proven fact. Instead, they represent the active edge of inquiry that could reshape understanding over time.

3. Hypothesis, Interpretation, and Speculation

The conceptual leap — the heart of this book. This is where measurable phenomena suggest a deeper mechanism and where love's coherence, resonance, and connection resemble perception, not mere emotion and where love behaves like a signal, not just a feeling.

These ideas are presented not as conclusions, but as openings — a framework for exploration, reflection, and lived experience. This distinction matters because the future of science is built where curiosity meets evidence.

This book stands on the foundation of what we know, builds upon what we are beginning to discover, and then steps consciously, transparently, into the territory of a working hypothesis. It does not ask the reader to believe. It asks the reader to consider and to feel into possibility, and to notice what resonates in their own direct experience.

Citations & Supporting Research By Chapter

Chapter 1:

"Sensory Receptors Are Responsive to a Single Type of Stimulus Energy." Neuro New Me, 2020. neuronewme.com

Chapter 2:

[1] Van Cappellen, P., Way, B.M., Isgett, S.F., & Fredrickson, B.L. (2016). Effects of oxytocin administration on spirituality and emotional responses to meditation. Social Cognitive and Affective Neuroscience, 11(10), 1579–1587.

[2] Baron Short, E., Kose, S., Mu, Q., Borckardt, J., Newberg, A., George, M.S., et al. (2010). Regional brain activation during meditation shows time and practice effects: an exploratory FMRI study. Evidence-Based Complementary and Alternative Medicine, 7, 121–127.

Strathearn, L., Fonagy, P., Amico, J., & Montague, P.R. (2009). "Adult Attachment Predicts Maternal Brain and Oxytocin Response to Infant Cues." Neuropsychopharmacology, 34(13), 2655–2666. (Avoidant attachment and altered oxytocin processing.)

Feldman, R. (2012). "Oxytocin and Social Affiliation in Humans." Hormones and Behavior, 61(3), 380–391.

Chapter 3:

Kowal, Magdalena, et al. "Love as a Commitment Device: Evidence from a Cross-Cultural Study." PNAS Nexus 3, no. 9 (2024).

Bode, Adam, et al. "Romantic Love as a Motivational State: A Cross-Cultural Study." Personality and Individual Differences (2025).

Fisher, Helen, Arthur Aron, and Lucy L. Brown. "Romantic Love: An fMRI Study of a Neural Mechanism for Mate Choice." The Journal of Comparative Neurology 493, no. 1 (2005): 58–62.

"Love and the Brain." Harvard Medical School Magazine, 2016.

Kosfeld, Michael, et al. "Oxytocin Increases Trust in Humans." Nature 435, no. 7042 (2005): 673–76.

Zimmerman, J. (1990). "Laying-on-of-Hands Healing and Therapeutic Touch: A Testable Theory." BEMI Currents, Journal of the Bio-Electro-Magnetics Institute, 2, 8–17. (SQUID magnetometer detection of pulsating biomagnetic fields from

healers' hands, frequency range 0.3–30 Hz, primary activity at 7–8 Hz.)
Seto, A., Kusaka, C., Nakazato, S., et al. (1992). "Detection of Extraordinary Large Bio-Magnetic Field Strength from Human Hand During External Qi Emission." Acupuncture & Electro-Therapeutics Research, 17(2), 75–94. (Japanese confirmation of healers' biomagnetic fields in QiGong, yoga, and meditation practitioners.)
Oschman, J.L. (2000). Energy Medicine: The Scientific Basis. London: Churchill Livingstone. (Comprehensive review of Zimmerman and Seto findings on biomagnetic healing frequencies.)

Chapter 4:

Craig, A. D. "Interoception: The Sense of the Physiological Condition of the Body." Nature Reviews Neuroscience 3, no. 8 (2002): 655–66.
Wang, Connie X., et al. "Transduction of the Geomagnetic Field as Evidenced from Alpha-band Activity in the Human Brain." eNeuro 6, no. 2 (2019).
Clery, Daniel. "Humans—Like Other Animals—May Sense Earth's Magnetic Field." Science News, March 18, 2019.
McCraty, Rollin, Mike Atkinson, William Tiller, et al. "The Effects of Emotions on Short-Term Power Spectrum Analysis of Heart Rate Variability." The American Journal of Cardiology 76, no. 14 (1995): 1089–93.
McCraty, Rollin. "Cardiac Coherence, Self-Regulation, Autonomic Stability, and Psychosocial Well-Being." Frontiers in Psychology 5 (2014): 1090.
McCraty, R. (2004). "The Energetic Heart: Bioelectromagnetic Communication Within and Between People." In: P.J. Rosch & M.S. Markov (Eds.), Clinical Applications of Bioelectromagnetic Medicine. New York: Marcel Dekker, pp. 541–562. (Heart's electromagnetic field approximately 60 times greater amplitude than brain waves; magnetic field more than 100 times stronger than the brain's; detectable several feet from the body.)

Chapter 5:

Kandel, Eric R., et al. Principles of Neural Science. 5th ed. New York: McGraw-Hill, 2013. (Sections on sensory receptors and transduction.)
"Sensory Receptors." In Anatomy & Physiology 2e. OpenStax, 2024.

Roper, Stephen D., and Alan C. Spector. "Encoding Taste: From Receptors to Perception." Comprehensive Physiology 11, no. 2 (2021): 1491–1524.

Chapter 6:

Aron, Arthur, et al. "Reward, Motivation, and Emotion Systems Associated with Early-Stage Intense Romantic Love." Journal of Neurophysiology 94, no. 1 (2005): 327–37.

Fisher, Helen, Arthur Aron, and Lucy L. Brown. "Romantic Love: An fMRI Study of a Neural Mechanism for Mate Choice." The Journal of Comparative Neurology 493, no. 1 (2005): 58–62.

Young, Larry J., and Zuoxin Wang. "The Neurobiology of Pair Bonding." Nature Neuroscience 7, no. 10 (2004): 1048–54.

Kosfeld, Michael, et al. "Oxytocin Increases Trust in Humans." Nature 435, no. 7042 (2005): 673–76.

Ito, Eiko, et al. "A Novel Role of Oxytocin: Oxytocin-Induced Well-Being in Humans." Journal of Physiological Sciences 69, no. 6 (2019): 717–27.

Hoehne, Katharina, et al. "Plasma Oxytocin Is Modulated by Mental Training, but Does Not Mediate Reduced Stress Reactivity." Psychoneuroendocrinology 139 (2022): 105708.

Osugo, Miriam, et al. "Striatal Dopamine D2/D3 Receptor Regulation of Human Reward Processing." Molecular Psychiatry 30, no. 1 (2025): 269–80.

Xu, Ming, et al. "Dopamine D3 Receptor Mutant Mice Exhibit Increased Behavioral Sensitivity to Dopamine Agonists." Neuron 19, no. 4 (1997): 837–48.

Armour, J.A. (1991). "Intrinsic Cardiac Neurons." Journal of Cardiovascular Electrophysiology, 2(4). (Original discovery of the heart's intrinsic nervous system containing approximately 40,000 neurons—the "heart brain"—capable of learning, remembering, and making functional decisions independently of the cranial brain.)

Armour, J.A. (2007). "The Little Brain on the Heart." Cleveland Clinic Journal of Medicine, 74(Suppl 1), S48–S51. (Comprehensive review of heart-brain neural communication, including the finding that the vagus nerve is approximately 80% afferent—carrying information from heart to brain, not the reverse.)

McCraty, R. (2004). "The Energetic Heart: Bioelectromagnetic Communication Within and Between People." In: P.J. Rosch & M.S. Markov (Eds.), Clinical Applications of Bioelectromagnetic

Medicine. New York: Marcel Dekker, pp. 541–562. (Heart's electrical field 60x greater amplitude than brain waves; magnetic field 100x stronger; detectable several feet from the body using SQUID magnetometers; field permeates every cell.)

McCraty, R., Atkinson, M., Tomasino, D., & Tiller, W.A. (1998). "The Electricity of Touch: Detection and Measurement of Cardiac Energy Exchange Between People." In: K.H. Pribram (Ed.), Brain and Values: Is a Biological Science of Values Possible. Mahwah, NJ: Lawrence Erlbaum Associates, pp. 359–379. (One person's heartbeat detected in another person's brainwaves during proximity and touch; mother's brainwaves synchronize to baby's heartbeat; signal amplifies approximately tenfold when holding hands.)

Kjaer, T.W., Bertelsen, C., Piccini, P., et al. (2002). "Increased Dopamine Tone During Meditation-Induced Change of Consciousness." Cognitive Brain Research, 13(2), 255–259. (First in vivo demonstration of increased endogenous dopamine release during meditation—65% increase in ventral striatum, measured via D2/D3 receptor binding, correlated with decreased readiness for action and EEG theta activity.)

Porges, S.W. (2011). The Polyvagal Theory: Neurophysiological Foundations of Emotions, Attachment, Communication, and Self-regulation. New York: W.W. Norton. (Neural pathways linking heart to brain; vagal tone and social engagement.)

Chapter 7:

Greene, Brian. The Fabric of the Cosmos: Space, Time, and the Texture of Reality. New York: Knopf, 2004. (Chapters on quantum fields and entanglement.)

Aspect, Alain, Philippe Grangier, and Gérard Roger. "Experimental Realization of Einstein-Podolsky-Rosen–Bohm Gedankenexperiment: A New Violation of Bell's Inequalities." Physical Review Letters 49, no. 2 (1982): 91–94.

Milonni, Peter W. The Quantum Vacuum: An Introduction to Quantum Electrodynamics. San Diego: Academic Press, 1994.

Brillante, Walter. "Zero-Point Energy and the Quantum Vacuum." Reviews of Modern Physics 57, no. 3 (1985): 783–836.

Engel, Gregory S., et al. "Evidence for Wavelike Energy Transfer through Quantum Coherence in Photosynthetic Systems." Nature 446, no. 7137 (2007): 782–86.

Panitchayangkoon, Gitt, et al. "Long-Lived Quantum Coherence in Photosynthetic Complexes at Physiological Temperature."

Proceedings of the National Academy of Sciences 108, no. 52 (2010): 20908–12.
Cha, Yuan, Christopher J. Murray, and Judith P. Klinman. "Hydrogen Tunneling in Enzyme Reactions." Science 243, no. 4896 (1989): 1325–1330.
Klinman, Judith P., and Amnon Kohen. "Hydrogen Tunneling Links Protein Dynamics to Enzyme Catalysis." Annual Review of Biochemistry 82 (2013): 471–496.
Scrutton, Nigel S., Michael J. Sutcliffe, et al. "Enzymology Takes a Quantum Leap Forward." Philosophical Transactions of the Royal Society B 361, no. 1472 (2006): 1375–1386.
Allemann, Rudolf K., and Nigel S. Scrutton, eds. Quantum Tunnelling in Enzyme-Catalysed Reactions. Cambridge: Royal Society of Chemistry, 2009.
Kohen, Amnon, and Judith P. Klinman. "Enzyme Dynamics and Hydrogen Tunnelling in a Thermophilic Alcohol Dehydrogenase." Nature 399 (1999): 496–499.
Hameroff, S., & Penrose, R. (2014). "Consciousness in the Universe: A Review of the 'Orch OR' Theory." Physics of Life Reviews, 11(1), 39–78. (Comprehensive review of the Orchestrated Objective Reduction model proposing that consciousness originates at the quantum level in microtubules within brain neurons. Includes responses to criticisms and integration of new evidence from quantum biology, including warm-temperature quantum vibrations confirmed experimentally.)
Penrose, R. (1989). The Emperor's New Mind: Concerning Computers, Minds, and the Laws of Physics. Oxford University Press. (Foundation for the argument that consciousness involves non-computable quantum processes.)
Hameroff, S., & Penrose, R. (1996). "Orchestrated Reduction of Quantum Coherence in Brain Microtubules: A Model for Consciousness." Mathematics and Computers in Simulation, 40(3–4), 453–480. (Original Orch OR proposal.)
Fröhlich, H. (1968). "Long-Range Coherence and Energy Storage in Biological Systems." International Journal of Quantum Chemistry, 2(5), 641–649. (Foundational paper predicting that biological systems possess coherent longitudinal electric modes; when energy is supplied above a threshold rate, a single mode becomes strongly excited and energy is stored in a highly ordered fashion resembling Bose-Einstein condensation—the theoretical basis for biological coherence.)

Laszlo, E. (2004). Science and the Akashic Field: An Integral Theory of Everything. Rochester, VT: Inner Traditions. (Theoretical framework proposing a universal information field underlying physical reality.)
Puthoff, H.E. (1989). "Gravity as a Zero-Point-Fluctuation Force." Physical Review A, 39(5), 2333–2342. (Zero-point field research and its implications for understanding vacuum energy.)

Chapter 8:

Backster, Cleve. "Evidence of a Primary Perception in Plant Life." International Journal of Parapsychology 10, no. 4 (1968): 329–48.
Horowitz, Kenneth, Donald Lewis, and Edgar Gasteiger. "Plant 'Primary Perception': Electrophysiological Unresponsiveness to Brine Shrimp Killing." Science 189, no. 4201 (1975): 478–80. (Failed replication of Backster's claims.)
McCraty, Rollin, et al. "The Effects of Emotions on Short-Term Power Spectrum Analysis of Heart Rate Variability." The American Journal of Cardiology 76, no. 14 (1995): 1089–93.
Goldstein, Pavel, Jane R. Weissman-Fogel, and Simone G. Shamay-Tsoory. "The Role of Touch in Regulating Inter-Partner Physiological Coupling During Pain." Scientific Reports 7 (2017): 3254.
Retallack, D. (1973). The Sound of Music and Plants. Santa Monica, CA: DeVorss & Company. (Experiments on plant growth responses to different types of music and sound.)
Vogel, M. (1974). "Man-Plant Communication." In: J. White & S. Krippner (Eds.), Future Science. New York: Anchor/Doubleday. (Early research on plant responses to human intention. Note: Primarily published through IBM internal reports rather than peer-reviewed journals.)
Popp, F.A., Nagl, W., Li, K.H., Scholz, W., Weingärtner, O., & Wolf, R. (1984). "Biophoton Emission: New Evidence for Coherence and DNA as Source." Cell Biophysics, 6, 33–51. (Discovery that all living systems emit a permanent ultra-weak photon flux of 1–100 photons/sec/cm^2 in the visible and UV range, displaying coherent properties; DNA proposed as primary source.)
Popp, F.A. (2003). "Properties of Biophotons and Their Theoretical Implications." Indian Journal of Experimental Biology, 41(5), 391–402. (Comprehensive review of biophoton coherence, intercellular communication, and biological regulation.)

Gagliano, M., Renton, M., Depczynski, M., & Mancuso, S. (2014). "Experience Teaches Plants to Learn Faster and Forget Slower in Environments Where It Matters." Oecologia, 175(1), 63–72. (Mimosa pudica can learn through habituation and retain learned behavior for weeks—evidence of plant memory without a nervous system.)

Simard, S.W., Perry, D.A., Jones, M.D., et al. (1997). "Net Transfer of Carbon Between Ectomycorrhizal Tree Species in the Field." Nature, 388, 579–582. (Foundational study on resource sharing between trees through underground mycorrhizal networks—the "wood wide web.")

Handlin, L., Hydbring-Sandberg, E., Nilsson, A., et al. (2011). "Short-Term Interaction Between Dogs and Their Owners: Effects on Oxytocin, Cortisol, Insulin and Heart Rate." Anthrozoös, 24(3), 301–315. (Physiological synchrony and oxytocin changes during dog-human interaction.)

Gehrke, E.K., Baldwin, A., & Schiltz, P.M. (2011). "Heart Rate Variability in Horses Engaged in Equine-Assisted Activities." Journal of Equine Veterinary Science, 31(5–6), 278–279. (Heart coherence between horses and humans.)

Fröhlich, H. (1968). "Long-Range Coherence and Energy Storage in Biological Systems." International Journal of Quantum Chemistry, 2(5), 641–649. (See also Chapter 7. Relevant here for biological coherence as a possible mechanism for intercellular and inter-organism communication.)

Chapter 9:

Aron, Arthur, et al. "Reward, Motivation, and Emotion Systems Associated with Early-Stage Intense Romantic Love." Journal of Neurophysiology 94, no. 1 (2005): 327–37.

McCraty, Rollin. "Cardiac Coherence, Self-Regulation, Autonomic Stability, and Psychosocial Well-Being." Frontiers in Psychology 5 (2014): 1090.

McCraty, Rollin, Mike Atkinson, and Dana Tomasino. "Science of the Heart." HeartMath Institute Monograph, 2015. (Chapter on psychophysiological coherence.)

Qaiser, Javeria, et al. "Shared Hearts and Minds: Physiological Synchrony During Empathy." Biological Psychology 181 (2023): 108545.

Lin, Dan, et al. "Emotion Contagion and Physiological Synchrony." Physiology & Behavior 266 (2024): 114–22.

Damasio, A.R. (1994). Descartes' Error: Emotion, Reason, and the Human Brain. New York: G.P. Putnam's Sons. (Foundational distinction between emotions—bodily action programs involving changes in heart rate, hormones, muscle tone, and facial expression—and feelings—the conscious mental experience of those bodily states. Introduced the somatic marker hypothesis.)

Damasio, A.R. (1999). The Feeling of What Happens: Body, Emotion and the Making of Consciousness. New York: Harcourt Brace. (Extended neuroscience of the emotions/feelings distinction and its role in consciousness.)

Damasio, A.R. (2003). Looking for Spinoza: Joy, Sorrow, and the Feeling Brain. New York: Harcourt. (Neural mapping of feelings and the body's role in generating conscious emotional experience.)

Ekman, P. (1992). "An Argument for Basic Emotions." Cognition and Emotion, 6(3–4), 169–200. (Foundational research identifying universal basic emotions across cultures.)

Chapter 10:

Lutz, Antoine, et al. "Long-Term Meditators Self-Induce High-Amplitude Gamma Synchrony During Mental Practice." Proceedings of the National Academy of Sciences 101, no. 46 (2004): 16369–73.

Qaiser, Javeria, et al. "Shared Hearts and Minds: Physiological Synchrony During Empathy." Biological Psychology 181 (2023): 108545.

Akimoto, Keiko, et al. "Effect of 528 Hz Music on the Endocrine System and Autonomic Nervous System." Health 10, no. 9 (2018): 1159–70.

Babayi, H., and G. Riazi. "The Effects of 528 Hz Sound Wave on Cell Viability and Oxidative Stress in Human Astrocytes." (Unpublished/limited-distribution study cited in secondary reviews.)

Emoto, Masaru. The Hidden Messages in Water. Hillsboro, OR: Beyond Words, 2004. (Original claims.)

Radin, Dean, et al. "Double-Blind Test of the Effects of Distant Intention on Water Crystal Formation." Explore 2, no. 5 (2006): 408–11.

Feldman, R., Magori-Cohen, R., Galili, G., Singer, M., & Louzoun, Y. (2011). "Mother and Infant Coordinate Heart Rhythms Through Episodes of Interaction Synchrony." Infant Behavior and

Development, 34(4), 569–577. (Mother-baby heartbeat synchronization during face-to-face interaction.)

Chapter 11:

McCraty, Rollin, and Dana Tomasino. "The Energetic Heart: Bioelectromagnetic Interactions Within and Between People." In Science of the Heart, HeartMath Institute, 2004/2015. (Chapter on energetic communication.)

Lutz, Antoine, et al. "Long-Term Meditators Self-Induce High-Amplitude Gamma Synchrony During Mental Practice." Proceedings of the National Academy of Sciences 101, no. 46 (2004): 16369–73.

"Study: Meditation Changes the Brain's Electrical Pattern." Center for Healthy Minds, University of Wisconsin–Madison, 2004.

Armour, J.A. (2008). "Potential Clinical Relevance of the 'Little Brain' on the Mammalian Heart." Experimental Physiology, 93(2), 165–176. (Review of the heart's intrinsic nervous system and its clinical implications.)

McCraty, R. (2015). Science of the Heart, Volume 2: Exploring the Role of the Heart in Human Performance. HeartMath Institute. (Comprehensive monograph on the heart's electromagnetic field, heart-brain communication, and energetic interactions between people.)

Zimmerman, J. (1990). "Laying-on-of-Hands Healing and Therapeutic Touch: A Testable Theory." BEMI Currents, Journal of the Bio-Electro-Magnetics Institute, 2, 8–17. (See also Chapter 3.)

Seto, A., et al. (1992). "Detection of Extraordinary Large Bio-Magnetic Field Strength from Human Hand During External Qi Emission." Acupuncture & Electro-Therapeutics Research, 17(2), 75–94. (See also Chapter 3.)

Targ, R., & Puthoff, H.E. (1974). "Information Transmission Under Conditions of Sensory Shielding." Nature, 251, 602–607. (Original publication of remote viewing research in a major peer-reviewed journal.)

Puthoff, H.E. (1996). "CIA-Initiated Remote Viewing Program at Stanford Research Institute." Journal of Scientific Exploration, 10(1), 63–76. (Declassified overview of the Stargate remote viewing program.)

Chapter 12:

Holt-Lunstad, Julianne, Timothy B. Smith, and J. Bradley Layton. "Social Relationships and Mortality Risk: A Meta-Analytic Review." PLoS Medicine 7, no. 7 (2010): e1000316.

McCraty, Rollin. "Cardiac Coherence, Self-Regulation, Autonomic Stability, and Psychosocial Well-Being." Frontiers in Psychology 5 (2014): 1090.

Kosfeld, Michael, et al. "Oxytocin Increases Trust in Humans." Nature 435, no. 7042 (2005): 673–76.

Doidge, N. (2007). The Brain That Changes Itself. New York: Viking. (Accessible overview of neuroplasticity research and the brain's capacity for structural reorganization through experience.)

Schlaug, G., Jäncke, L., Huang, Y., & Steinmetz, H. (1995). "In Vivo Evidence of Structural Brain Asymmetry in Musicians." Science, 267(5198), 699–701. (Musicians' auditory cortices are significantly enlarged compared to non-musicians—evidence that repeated practice physically reshapes brain structure.)

Kjaer, T.W., et al. (2002). "Increased Dopamine Tone During Meditation-Induced Change of Consciousness." Cognitive Brain Research, 13(2), 255–259. (See also Chapter 6. Relevant here for dopamine's role in meditation-induced states of transcendence and evolutionary implications of reward-system activation through spiritual practice.)

Newberg, A.B., et al. (2018). "Effect of a One-Week Spiritual Retreat on Dopamine and Serotonin Transporter Binding: A Preliminary Study." Religion, Brain & Behavior, 8(3). (Significant decreases in dopamine transporter binding in basal ganglia and serotonin transporter binding in midbrain following a 7-day retreat—neurochemical evidence of lasting changes from intensive spiritual practice.)

Chapter 13:

Chapter 13 draws upon research cited in earlier chapters, particularly the HeartMath research on heart coherence from Chapter 6, the neuroplasticity findings from Chapter 12, and the quantum biology foundations from Chapter 7. No additional unique citations required.

Made in the USA
Monee, IL
04 November 2026

67385337R00115